AN ILLUSTRATED ESSAY ON THE NOCTUIDAE of NORTH AMERICA with A COLONY OF BUTTERFLIES

BY

A. R. GROTE

1882

WITH A BIO-BIBLIOGRAPHICAL FOREWORD

BY

R. S. WILKINSON

Ph.D., F.L.S., F.R.E.S.

CLASSEY PUBLICATIONS

1971

E. W. CLASSEY LTD.

353, HANWORTH ROAD, HAMPTON

MIDDLESEX. ENGLAND

Reprinted E.W. CLASSEY LTD, 353, Hanworth Road,
Hampton, Middlesex, England. 1970.

1 SBN 0 900848 30 8

REPRINTED IN GERMANY
Hain–Meisenheim
ECM.

TO

Dr. A. S. Packard, Jun.

CONTENTS.

FOREWORD

Augustus Radcliffe Grote (1841–1903), author of *An Illustrated Essay on the Noctuidae of North America,* was one of the most significant nineteenth-century American entomologists. Despite his many contributions, an extensive account of his life has never been written, and his biographical data are scattered through numerous nineteenth and early twentieth-century sources.[1]

Grote was born in Aigburth, near Liverpool, on 7 February 1841. His father, Friedrich Rudolf Grote, was a German originally from Danzig, a descendant of the Dutch theologian Hugo Grotius; the family name had been Grohte until Friedrich became a British citizen. Grote's mother was Anna Radcliffe, the daughter of a Welsh ironmaster. In a brief autobiographical account communicated in the 1880s to William Saunders, editor of the *Canadian Entomologist,* Grote recalled that he came to the United States at the age of seven, following his parents who had emigrated in the previous year. His father purchased an extensive farm on Staten Island, New York, and there Grote quickly developed a youthful interest in insects.

While a student in Joseph Deghuée's school at Brooklyn, Grote met several other incipient entomologists. One was Edward Graef, who wrote much later about their association. The boys began their study of Lepidoptera about 1854, due to their reading of Graef's copy of J. W. Meigen's *Handbuch für Schmetterlingsliebhaber* (Aix-la-Chapelle, 1827). Grote, Graef and Frederick Tepper formed a considerable collection of insects in cigar-boxes, and eventually made the acquaintance of such noted Brooklyn collectors as John Akhurst and Stephen Calverley.[2]

In a passage on early dawn collecting written in 1886, Grote recalled his intensely formative experiences as a youth on Staten Island, where he roamed the relatively unspoiled lanes and sensed a contact with nature which would have satisfied his Transcendentalist contemporaries.[3] As he grew older, his intensive collecting of Lepidoptera and Coleoptera ceased. He explained the reason in 1882: "These insects were part of a Universe with Stars and Suns. I could not understand the life I was taking; and then I felt the grief that arises when we become conscious of the role played by Destruction. I abandoned collecting insects, even walking carefully so as not to bruise the golden-rods and purple asters that fringed my path ... and joined the Battle of the Synonyms instead."[4]

Grote later renounced this vow due to the advent of relatively humane killing devices, and his realization that entomologists reduce total populations only infinitesimally. Yet his experiences on Staten Island determined the course of his life, and his intense feeling for nature was always evident in his writings. It has been said that had Henry David Thoreau specialized in the Lepidoptera, he might well have written an account similar in essence to Grote's "Collecting Noctuidae by Lake Erie" (1895), one of the most literary examples of entomological literature.[5]

Grote's father invested heavily in real estate and the Staten Island Railway, experiencing serious financial loss in the panic of 1857. This interrupted young Grote's preparation for Harvard, and although (according to his own account) he eventually managed to study on the Continent, the present author has been able to trace only the degree that Grote frequently claimed, the A. M. This was honorary, granted in 1874 by Lafayette College, Easton, Pennsylvania.[6]

Grote's intensive study of the Noctuidae began in 1859–60,[7] and his earliest papers on the Lepidoptera were published in 1862. A recruit to the "Battle of the Synonyms", he was elected a corresponding member of the Entomological Society of Philadelphia (now the American Entomological Society) in that year.[8] His first species description seems to have been that of the noctuid *Heliothis umbrosus* (now *umbrosa,* a form of *H. obsoleta* [Fabr.]), in the first of his series of "Additions to the Catalogue of U. S. Lepidoptera."[9] Grote later recalled that when he began his work, "there were hardly any species [of the Noctuidae] named in any North American collection beyond the few mentioned by Harris.[10] The species described by Guenee in Europe were unidentified. These had to be made out and gradually brought into natural genera. Mr. [Francis] Walker's species were wholly unknown and, worse than this, were undeterminable from the descriptions published."[11]

The situation was not nearly as desolate as Grote remembered it, but there was much room for diligent work. The earliest papers in his massive taxonomic assault were prepared at Buffalo, New York, where Grote joined the Society of Natural Sciences in 1862 and became a member of its Board of Managers.[12] Here he began his long collaboration with Coleman Townsend Robinson (1838–72), who had been instrumental in founding the Buffalo society in the previous year.[13] Robinson was born in Putnam County, New York. He studied at the University of Berlin, and upon his return in 1861 he engaged in business as a stockbroker in Wall Street. He became head of Robinson, Cox and Co., and had a net worth of a million and a half dollars when he retired at an early age. He died as the result of a fall from his carriage.

Robinson's avocation was entomology, and most of his publications concerned the Microlepidoptera.[14] The Grote-Robinson papers began appearing in 1865, and eventually dealt with a wide range of the Lepidoptera. Their first important contribution was a catalogue of the North American Sphingidae (1865)[15], and the Grote-Robinson series of "Descriptions of American Lepidoptera," which contained most of their new species, was cut short by Robinson's death.[16] While preparing this series, the two friends visited the British Museum to examine Francis Walker's types, and an important result was their joint critique (1868) of the American species described by Walker in the *List of the Speci-*

mens of Lepidopterous Insects in the ... British Museum (London, 1854–66).[17] The Grote-Robinson *List of the Lepidoptera of North America,* the first thorough effort of the kind since J. G. Morris' pioneer *Catalogue of the Described Lepidoptera of North America* (Washington, 1860), was left unfinished at Robinson's death; the first part was published at Philadelphia in 1868.

During his first residence in Buffalo, Grote curated the entomological collections of the Society of Natural Sciences, and served as one of the editors of the *Practical Entomologist* (1865–66), a free monthly bulletin for farmers published by the Entomological Society of Philadelphia. In 1868 he was elected an honorary member of the Entomological Society of Canada (now of Ontario), and he eventually contributed a very large number of papers and notes to the *Canadian Entomologist.*

The details of Grote's first marriage are still unclear. He was wed in 1869 or 1870 to Julia (also given as Tulia) Blair,[18] and there were two children, Anna and Julia. The family lived in Demopolis, Alabama, where Grote undertook both taxonomic and economic studies of Southern Lepidoptera. In a series of papers on the "cotton-worm," the noctuid *Alabama argillacea* (Hbn.), Grote accepted the theory (suggested as early as 1846) that the insect migrated northward yearly from more southern areas and did not complete its life-cycle in the United States. The noted American entomologist Charles Valentine Riley claimed that *argillacea* did indeed overwinter successfully, which is now known to be true, and a controversy began between Grote and Riley which would have unhappy results.

When Grote travelled to Washington in 1874 to urge an appropriation for the investigation of cotton pests, he met with little encouragement. When in 1877 the U. S. Government created an Entomological Commission to study insects injurious to agriculture, Riley was appointed chief, and Grote later admitted his keen disappointment at not being named to the group. He felt that this was due to Riley's animosity; indeed, beginning with the "cotton-worm" controversy, he had made use of every opportunity to criticize Riley's work. The two remained enemies for years.

Another of Grote's adversaries was Ferdinand Heinrich Herman Strecker, sculptor, entomologist and author of *Lepidoptera, Rhopaloceres and Heteroceres* (Reading, Pa., 1872–1900). The long feud between Strecker and Grote is the most colourful in American entomology. Neither could please the other in the identification of species and naming of taxa. Riley was a mild adversary, but much of Strecker's *Lepidoptera* series was peppered with stinging attacks on Grote. Strecker's was the greater wit, and although such a performance could hardly be tolerated today, the reputation of neither of these great taxonomists has been diminished by their unfortunate quarrel. Strecker's favourite saw, "Time at last sets all things even," was apparently correct.

Grote's first major paper on the genus *Catocala* was written during his Alabama residence. In 1872 he revealed that he had lost in transportation an extensive monograph on the genus in America, which could not be reconstructed as the material on which it was based had been returned to various collections. In the pioneer summary which he did publish, he recognized 58 species, 16 of which he had described, some in conjunction with Robinson. Grote continued to give special attention to the *Catocala,* and 22 of his species are still valid, as are two described by Grote and Robinson.[19]

Grote's wife died during the birth of their second child, and he returned to the Buffalo Society at his own request in January, 1873, accepting the positions of librarian and museum director at a salary of $ 500 yearly. He began publication of the Society's *Bulletin,* and contributed a large percentage of the papers himself. These were mainly on the Lepidoptera, although there were articles on other topics. Among Grote's important publications in the *Bulletin* were his "List of the Noctuidae of

North America" (1874), and catalogues of the Sphingidae and Zygaenidae (1873). During these very busy years he contributed papers to a number of journals, including the *Proceedings* of the American Philosophical Society, the *Annals* of the New York Lyceum of Natural History, the *Proceedings* of the Academy of Natural Sciences of Philadelphia, the *Transactions* of the American Entomological Society, the Stettin *Entomologische Zeitung,* the *American Naturalist, Scientific American* and *Popular Science Monthly.* His contributions to the *Canadian Entomologist* continued, and his papers appeared in the *Annual Report* of the Peabody Academy of Sciences and the *Bulletin* of the U.S. Geological and Geographical Survey of the Territories.

In 1875 Grote published his *Check List of the Noctuidae of America, north of Mexico,* and he was vice-president of Section B of the American Association for the Advancement of Science in 1878. Grote founded and edited a new journal in 1879–80, the *North American Entomologist,* and contributed papers to most of the monthly numbers. His general appearance in the 1870s was recalled much later by a friend; Grote was "dark, slight in build, and of a nervous, mercurial temperament ... a man of general culture, quite a poet, and devoted to music." Grote was a "delightful talker," and when he was not "occupied in identifying or describing some new species, he was writing a lyric, composing an opera, or recasting theology."[20] (An early convert to Darwinism, his 'theological' speculations would soon become apparent.) Grote was organist of an Episcopal church in Buffalo, and his musical production occupied its share of his leisure time; one obituary mentions "48 musikalische Opera, deren Nennung hier zu weit führen würde."[21]

Due to the death of his father, Grote decided to leave Buffalo and return to Staten Island. He resigned his position at the Society of Natural Sciences in April 1880, and in July he relinquished his editorship of the *North American Entomologist,* which was not continued. He read several papers at the founding meeting of the New York Entomological Club in December, was elected President, and served on the publication committee of the Club's journal *Papilio,* edited by the noted lepidopterist Herny Edwards. Grote's contributions to *Papilio* were numerous until Edwards resigned his office in 1884 and the journal was moved to Philadelphia.

Grote's first contribution to the nineteenth-century conflict between science and religion, *Genesis I–II: An Essay on the Bible Narrative of Creation,* was printed at New York in 1880. He argued that those who were attempting to reconcile modern scientific discoveries with Genesis were misguided; the Bible obviously contradicted gradual development, and could no longer be accepted as literal truth, although certainly it was once meant to be so. The first two chapters of Genesis were printed in transliterated Hebrew and English in parallel columns, and these were followed by chapters which fixed the approximate dates and sources of the sometimes contradictory accounts making up the creation-myth, considering the testimony of archaeology and comparative mythology.

Grote gave examples of the unscientific nature of Genesis and suggested that it was written by a people who had a very low level of scientific knowledge. His argument was "not that the teachings of Christ are to be rejected, or the morality of The Hebrew Bible to be condemned, but that we are to correct our views as to the way in which existing plants and animals, including man, came to be what they are today."[22] *Genesis I–II* enjoyed at least a limited popularity; the small and poorly printed first edition was sold out before the spring of 1881, and a revised, enlarged and more attractive edition appeared in May.

Both editions are very scarce,[23] as is Grote's second book on the subject, *The New Infidelity* (New York, 1881). In it he argued for a logical reconciliation of religion and science, with each valued according to its merits. His 'new infidelity' was the scientific view, which rejected the supernaturalism of religion while understanding it in the light of comparative and historical studies. Religion was a human institution, but had helped in the progress of mankind, and it should be improved rather than abolished.[24]

The New Infidelity appeared in November, and in the following January a collected edition of Grote's poems was printed at London. *Rip van Winkle: A Sun Myth and Other Poems* (1882) included compositions which had been published in the *Atlantic Monthly*, the *Buffalo Courier*, *Evolution* and other periodicals. The principal poem, "Rip van Winkle, a Sun Myth," had been written in 1874–5.[25] Grote's other important separate publications in 1882 were his partial *New Check List of American Moths*, printed at New York, and the *Illustrated Essay on the Noctuidae of North America*. Arrangements for publication of his poems and the *Illustrated Essay* had been made during his visit to England in November, 1881.

Ostensibly this journey was made for reasons of health, but Grote's chief intent was to transport his well-known collection of Lepidoptera to London for possible sale. He had been in poor financial condition for some time, and his friend Graef wrote of debts which, as a matter of honor, Grote "thought he must pay without delay."[26] His collection, which had been returned to him at his request by the Buffalo Society of Natural Sciences, was offered for sale in the United States at a price of $ 5000. No American institution would meet the cost, and in February 1881 Grote suggested a price of £ 850 to the British Museum. This was thought too high, and Grote reluctantly lowered the amount to £ 700.

Upon examining the insects in London, Dr. Albert Günther strongly recommended that the Trustees should approve the purchase. There were approximately 2600 species represented by 6500 specimens, of which at least 1100 were types. Günther noted that their acquisition "would be especially important to the British Museum which is so extremely poor in North American Lepidoptera that scarcely any of the specimens in Mr. Grote's collection would be duplicates."[27] Purchase was approved, and loss of the collection was generally lamented by American lepidopterists; Henry Edwards wrote that had it "been composed of old engravings, broken china, or even of minerals and fossils," an American purchaser would have readily been found."[28]

Grote himself was disappointed with the results of the transaction; he later told William Saunders that failure to obtain a position on the British Museum staff was keenly discouraging. The financial difficulties were only temporarily alleviated. In February 1883 he wrote to Lord Walsingham that he was in the hands of money-lenders, and would be forced to sell the books and specimens still in his hands unless immediate aid was forthcoming; he was tired of life, and his only comfort was the companionship of his two young daughters.[29]

In poor health and still suffering from debt, Grote left the United States in 1884 and took up residence in Bremen, where he married Gesa Maria ("Minna") Ruyter, daughter of J. Ludwig Ruyter, a prominent politician and wholesale dealer in tobacco. There were four more children; the famous internist Louis Radcliffe (1886–1960)[30], Lotty, Theodore and Mathilde. Grote's fortunes revived in the period following his second marriage, and in 1886 his publications poured forth again after a two-year period during which he wrote very little. The scope of his investigations broadened visibly in Germany. He had hardly confined himself to taxonomical problems in the past, but now research into other areas occupied much more of his time.

In his later papers Grote investigated a wide variety of subjects, including evolution of families within the Lepidoptera, larval and adult ornamentation, larval characters, neuration, mimicry, sounds emitted by Lepidoptera, defence mechanisms and others. He developed a new interest in the classification of the Saturniidae, but his old preferences were not forgotten. His *Hawk Moths of North America* was printed at Bremen in 1886, and in 1890 he published the first part of his *Revised Check List of the North American Noctuidae.* There was an eight-part series comparing the Noctuidae of North America and Europe (1889–90) in the *Canadian Entomologist,* which continued to receive further papers until Grote's death.

Although he partially recovered his amazing productivity in the German period of his life, the range of journals to which Grote communicated his papers was somewhat narrowed. A notable addition was the *Entomologist's Record*; articles appeared in *Entomologica Americana* and the *Journal* of the New York Entomological Society, as well as the *Insektenborse* and other German journals. In 1895 Grote and his family removed to Hildesheim, where his catholicity of interests led him to become curator of the Roemer-Museum. He vitalized the Museum's publication program much as he had done in Buffalo, and it will surely confuse computer-bibliographers of the future to discover that most of Grote's *Mitt. Roemer-Museum Hildesheim* was devoted not to antiquities but to the Lepidoptera.

Grote died of endocarditis on 12 September 1903, ending a remarkably productive life. From promising beginnings in 1862, he set a course which has been practically unparalleled in the study of American Lepidoptera. He was the first American investigator to study the Noctuidae in real depth, and he successfully transferred the study of U.S. Noctuidae to America. Aside from the great number of species to be named and genera to be defined, he turned his attention (as did his collaborator Robinson) to the insufficient or confusing species descriptions of European investigators. Grote helped to further the idea that American species were essentially different from similar European ones (although he recognized a number of circumpolar species).

Although he investigated most areas of lepidopterology, Grote's greatest contribution was undoubtedly the description of a vast number of American species. Almost 1250 of his taxa are recorded (many as synonyms) in the currently-used McDunnough and dos Passos check lists, and over 140 taxa are credited to Grote and Robinson.[31] Essig's very useful but sometimes inaccurate biographical compendium states that Grote "published about 100 papers on Lepidoptera,"[32] yet Herman Strecker listed over 160 of Grote's publications as early as 1878.[33] Later compilers have raised the total to over 600[34], but no single bibliographer has risked offering a 'complete' list of Grote's scattered papers. Future investigators will have to add non-entomological publications and popular articles, for these are hardly without importance. Grote was neither a Fiske nor a Huxley, but he contributed to the popular science of his day, and if his efforts in the conflict between science and Scripture failed to gain a wide audience, they were at least timely and sound.

Grote negotiated with Van Voorst for publication of the *Illustrated Essay* during the 1881 visit to London. A re-appraisal of the Walker types was made at this time, using for comparison Grote's own specimens, which had been brought for possible sale to the British Museum. The "Notes on Mr. Walker's Types" contained in the volume resulted from this examination. Grote's hope that the British Museum would undertake the laborious work still to be done on the Walker specimens was eventually realised; Sir George Hampson's monumental *Catalogue* began its appearance sixteen years later.

Grote's preface and general essay on the Noctuidae furnished a useful summary of the history and current status of research on the family, especially in America. Presented in a popular style, little of the material was original, but it was hardly meant to be so. The attack on C.V. Riley in the preface was certainly out of place in such a publication, even when Grote's long-standing feud with the U.S. Entomologist is considered, and the author's propensity to call attention to his own contributions is evident throughout the work. Despite these possible shortcomings, Grote's call for "serious biological work" on the North American Noctuidae was well taken, as was his timely (if unoriginal) insistence that descriptions on the basis of general appearance were superficial and incorrect. Grote joined others in suggesting more frequent use of the microscope in entomology, and his caution regarding the wholesale use of poisons to combat economic pests was prophetic.

While Grote recognized approximately 1460 North American species in the family, he suggested that "about two thousand species of Noctuidae will be found in our territory." McDunnough (1939) listed 2693 species of North American Phalaenidae,[35] and the forthcoming *Moths of America North of Mexico* by Ferguson, Franclemont and Hodges will undoubtedly revise this figure. Admittedly, forty-one of the forty-five "Specimens of North-American Noctuidae" in the *Illustrated Essay* were Grote's species, but his discussions of them and the excellent (if somewhat highly coloured) plates, prepared in London from Grote's specimens by A.H. Searle, justify their inclusion in a book which after all was Grote's own.

"A Colony of Butterflies" first appeared in the *American Naturalist* for March, 1876,[36] and Grote revised it only slightly for inclusion in this volume. His interest in glaciation and its effect on insects also resulted in a series "On the Insect Fauna of the White Mountains" in *Psyche* (1875–76), and his paper on "The Effect of the Glacial Epoch upon the Distribution of Insects in North America," read before the American Association for the Advancement of Science in 1876.[37] "A Colony of Butterflies" was dedicated by Grote to the pioneer American entomologist Thomas Say, who named the White Mountain Butterfly (*Oeneis melissa semidea*) as *Hipparchia semidea* in 1828.[38] Statements in Grote's various papers show that he was not sure of the taxonomic relation of the *semidea* colony to other *Oeneis* populations, but the situation has been clarified with the assignment of *semidea* to subspecific rank.

Grote's call for protection of the White Mountain Butterfly was a pioneer American effort in the conservation of Lepidoptera. His request for legal protection was not heeded, but at least in this case, the results have been favourable. Due to its isolation, habits, and the good sense of those collectors who do visit Mount Washington, the colony still thrives, and has prospered to the extent that a leading American lepidopterist has recently suggested Mount Washington in early July as "one place where anyone can catch an *Oeneis*."[39] Unless circumstances change, *semidea* appears to be safe.

Grote's *Illustrated Essay* was well received upon its appearance in 1882. Henry Edwards described it in *Papilio* as "a notable contribution to N. Am. Entomological literature," which was perhaps too flattering an estimate,[40] but the little book has since held its own. Understandably, C. V. Riley answered Grote's criticism with a totally unfavourable review, which partially answered the charges but dealt rather unfairly with the rest of the book.[41]

The *Illustrated Essay* is now a very uncommon title; a brief census of American libraries reveals its absence from some of the more extensive collections of entomological works. As natural history booksellers will testify, the demand far exceeds the few examples which occasionally appear for sale, and this reprint edition will surely be welcome to many.

Ronald S. Wilkinson

REFERENCES

1 The most lengthy biographical notice is an obituary by C.J.D. Bethune in the Entomol. Soc. of Ontario *Annual Report* XXXIV (Toronto, 1904), 109–112. This is largely based on a manuscript written by Grote himself, but there are unfortunate misprints and other errors which have been perpetuated by later writers who have used Bethune as their source.

2 E. L. Graef, "Some Early Brooklyn Entomologists," *Bull.* Brooklyn Entomol. Soc. IX (1914), 47–56. Graef discusses Grote's youthful entomological activities in some detail.

3 A. R. Grote, *The Hawk Moths of North America* (Bremen, 1886), 10–12.

4 A. R. Grote, *New Check List of North American Moths* (New York, 1882), 5–6.

5 Grote's essay was printed in *Entomol. Rec.* VI (1895), 97–101, 121–125.

6 The information was kindly furnished by Margaret A. Schenck, Secretary to the Registrar, Lafayette College.

7 A. R. Grote, *Revised Check List of the North American Noctuidae* (Bremen, 1890), 46.

8 On 14 April 1862; *Proc.* Entomol. Soc. Phila. I (1862), 186.

9 *Proc.* Entomol. Soc. Phila. I (1862), 218–219.

10 T. W. Harris (1795–1856), Harvard College librarian and pioneer American economic entomologist, author of *A Report on the Insects of Massachusetts Injurious to Vegetation* (1841), which, retitled, went through numerous editions.

11 A. R. Grote, *Revised Check List of the North American Noctuidae* (Bremen, 1890), 46.

12 I am grateful to Dr. Virginia Cummings, Acting Director of the Buffalo Society of Natural Sciences, for much information concerning Grote.

13 Henry R. Howland, "The Buffalo Society of Natural Sciences: Historical Sketch," *Bull.* Buffalo Soc. Nat. Sci. VIII (1907), 4.

14 There is a bibliography by Grote, "List of the Writings of the late Coleman Townsend Robinson," *Canad. Entomol.* IV (1872), 109–111. An obituary by C. J. S. Bethune is in the same issue, 118.

15 A. R. Grote and C. T. Robinson, "A Synonymic Catalogue of North American Sphingidae, with Notes and Descriptions," *Proc.* Entomol. Soc. Phila. V (1865), 149–193.

16 A. R. Grote, "A Supplement to the 'Descriptions of American Lepidoptera'," *Trans.* Amer. Entomol. Soc. IV (1873), 425–435.

17 A. R. Grote and C. T. Robinson, "Notes on the North American Lepidoptera in the British Museum and described by Mr.

Francis Walker," *Trans.* Amer. Entomol. Soc. II (1868), 67–88.

18 The legal record has not been located. The Bethune obituary, supposedly from Grote's own account, gives the date as 1880, but this is evidently a misprint for 1870, as Bethune later states that Grote returned to Buffalo in 1873 as a result of his wife's death. It appears from other sources that Grote's years in Alabama coincided with his first marriage. In his *Dictionary of American Biography* account, L. O. Howard gives the dates of Grote's first marriage as 1880–1883. The *National Cyclopaedia of American Biography* suggests 1869 as the wedding date.

19 A. R. Grote, "On the North American Species of Catocala," *Trans.* Amer. Entomol. Soc. IV (1872), 1–20.

20 J. G. Milburn, "Recollections of A. R. Grote," *Entomol. News* XXIV (1913), 183.

21 W. Bode, "Prof. A. Radcliffe Grote," *Allgemeine Zeitschrift für Entomologie* IX (1904), 6.

22 A. R. Grote, *Genesis I–II: An Essay on the Bible Narrative of Creation* (New York, 1880), 7.

23 The Widener Library, Harvard University, has both editions. Both were printed by Asa K. Butts. The second edition has an added chapter on "The Bible and the Koran," and the Hebrew transcription is slightly revised.

24 The book was published by G. P. Putnam's Sons in New York. It was later translated into German. Grote confided to William Saunders that the book "met with warm praise and severe criticism, but enjoyed only a small sale."

25 Five hundred copies of *Rip van Winkle* were printed by Kegan Paul, Trench & Co.

26 Graef, *op. cit.*, 53.

27 Report of Albert Günther to the Trustees of the British Museum, 1 Dec. 1881; British Museum of Natural History.

28 Henry Edwards, "Mr. A. R. Grote's Collection," *Papilio* II (1882), 82. Grote or Grote-Robinson types are also in the American Museum of Natural History, the Academy of Natural Sciences (Philadelphia), the University of Kansas and perhaps other collections.

29 A. R. Grote to Lord Walsingham, 7 February 1883; British Museum of Natural History.

30 There is a biographical sketch of Louis R. Grote by H. U. Schulz in *Neue Deutsche Biographie* VII (Berlin, 1966), 163–164.

31 J. McDunnough, "Check List of the Lepidoptera of Canada and the United States of America," Part I, *Mem.* S. Calif. Acad. Sci. I (1938); II (1939); C. F. dos Passos, "A Synonymic List of the Nearctic Rhopalocera," *Mem.* Lepid. Soc. I (1964).

32 E. O. Essig, *A History of Entomology* (New York, 1931), 640.

33 F. H. H. Strecker, *Butterflies and Moths of North America* (Reading, Pa., 1878), 232–237.

34 W. Horn and S. Schenkling, *Index Litteraturae Entomologicae ... bis inklusive 1863* II (Berlin, 1928), 465; W. Derksen and U. Scheiding-Göllner, *Index Litteraturae Entomologicae 1864–1900* II (Berlin, 1965), 212–221.

35 McDunnough, *op. cit.,* I, 54–131.

36 "A Colony of Butterflies," *American Naturalist* X (1876), 129–132.

37 The White Mountain series appeared in *Psyche* I (1875), 76–77, 99–100; (1876), 131. The AAAS address was first printed in *Proc.* Amer. Ass. Adv. Sci. XXIV (1875), sec. B, 222–226; it was reprinted in other journals in the U.S.A. and England.

38 Thomas Say, *American Entomology* III (Philadelphia, 1828), pl. 50 and text. The classic discussion of *semidea* is that of Samuel Scudder in *The Butterflies of the Eastern United States and Canada* (Cambridge, Mass., 1889), 134–148.

39 Alexander B. Klots, *A Field Guide to the Butterflies* (Boston, 1960), 75.

40 Henry Edwards, "Book Notices," *Papilio* II (1882), 146.

41 C. V. Riley, "Notice of an 'Illustrated Essay on the Noctuidae of North America'," *Bull.* Brooklyn Entomol. Soc. V (1883), 77–79. The charge that the work was unoriginal is partially true but meaningless. Riley pointed out that both the facts and the theory contained in the "Colony" had been set forth previously by A. S. Packard and S. H. Scudder. It is true that Grote did not give these investigators credit in the "Colony", but in his earlier AAAS address on the results of glaciation he had cited both of the papers mentioned by Riley. Both Grote and Riley seem to have overlooked an earlier article by Scudder, "Remarks on Some Characteristics of the Insect-Fauna of the White Mountains, New Hampshire," *Bost. J. Nat. Hist.* VII (1863),

612–631. Grote admitted in the *Illustrated Essay* that his discussion of the cotton-worm was taken from his contribution to the Alabama *Geological Report* [for 1875], and Riley's observation that the paragraph committing Grote to the annual dying-out of the species in the United States had been omitted in the *Illustrated Essay* would seem to show Grote's good sense in the light of further evidence.

PREFACE.

At the commencement of this Essay I must draw attention to the fact that, as yet, no serious biological work has been attempted with the North-American Noctuidæ. The internal structure is yet a mystery to us, or at best we can judge it from Mr. Burgess's admirable work on *Danaus*, or get some notion of it from the observations brought together in Dr. Packard's 'Guide to the Study of Insects.' Some work, not perhaps so extended, but in the line of Professor Huxley's exhaustive volume on the Crawfish, is needed on any of our common species of Noctuidæ. This Essay deals with the external structure, as to which I do not find much notice taken by the older English lepidopterists, except by Stephens, whose discriminations of genera I find very good and anticipatory of the characters afterwards used by Lederer. I find, also, in Hübner's 'Verzeichniss' (and here, I fear, I shall hardly convince Mr. W. H. Edwards) evidence that the author had made some examinations of characters, or else some very extraordinary guesses, as, for instance, in the association of the genera under the *Bombycidæ*, where he was clearly ahead of his time. The genera, as established, are dependent on comparative characters or on details of absolute structure, these latter only to be observed by the microscope. The latter class are of the most value; yet, for convenience, some of the former may be considered valid. I discuss, therefore, in this Essay merely the external characters, the sorting of individuals into "species," and the literature of the North-American Noctuidæ. The method of classifying by mere appearance can lead to no scientific result; yet, *au fond*, this is the method employed by English lepidopterists, of whom the extreme example was Mr. Walker, and the best, perhaps, Mr. Doubleday. There must be a radical change in this procedure, of which we in the United States are the legatees, and show that we are, when we "lump" the genera allied to *Hesperia* or *Eudryas*. The microscope must be thoroughly used at every step, and then the dissecting-needle. We are yet employed merely in getting our cabinets into order; and the real results we are to obtain, the connexions of the Noctuidæ with the chain of living organisms, are ahead of us in time.

The study of Nature and the collection and examination of objects of Natural History is a favourite occupation of our race. For those who have patience and some manual dexterity, the formation of a collection of Butterflies and Moths will give pleasure and instruction. It is better if it be undertaken in connexion with a study of the structure and habits of the insects, viewing them in their relation to the rest of animated Nature and their immediate surroundings. It is not difficult to become acquainted with the external appearance of the different parts of the body in the Lepidoptera, although the hair and scales covering the body and wings must be removed in order that the shape of the pieces making up the thorax and head, and the course of the veins, be clearly observed. An excellent method of taking the colouring-matter out of the scales of the wings, rendering them perfectly transparent, has been discovered by Mr. George Dimmock; and my friend Professor C. H. Fernald has explained the method employed by him in mounting the prepared wings as microscopic objects with such success. A knowledge of the structure of the legs is of importance in the classification of the Noctuidæ, in order to locate the species generically; but this can be observed with a good lens (I have used a half-inch on a binocular stand), and generally without any denuding, although the armature of the front tibiæ is sometimes concealed by the vestiture.

The growth of the Moths may be divided into the several stages of egg, caterpillar, chrysalis, and imago or perfect insect, as these are severally easily observed by us. Yet the life of the individual very gradually proceeds, despite the apparent suddenness of the transformations it undergoes. From another point of view we may consider the life of the Moth as falling into two periods—its immature existence, and the final mature state in which it is able to reproduce its kind. The egg, caterpillar, and chrysalis mark epochs in its immature condition, the two latter stages not being as completely defined in some of the other Orders of Insects as in the Lepidoptera.

The Moths belong to the ringed or jointed type of animals, and, theoretically, the individuals may be regarded as compound. As we descend in rank in this type of structure, we find that there is a tendency towards independence on the part of the rings out of which the body is composed. The lower Worms are rather colonies of individual rings, each segment being furnished with organs of locomotion, respiration, digestion, and reproduction variously modified. In the Moths (in which, with other six-footed insects, the separation of the functions is carried to its highest extent) the processes of respiration, digestion, and reproduction are carried on by organs situated in the abdominal region. The middle region supports the legs and wings, and the rings have become curiously welded and arched in order to give firmness and room for muscular development for the organs of locomotion; the anterior part of the body, or "head," is provided with a mouth and "tongue," and carries perceptive organs, by means of which the insect places itself in the most favourable conditions for its existence. All the details in the life-history of any one species supply material for our observation of the way in which insect-life adapts itself to the environment. And it is here that the study of Entomology becomes of real interest and is relieved from the childish aspect, which it else wears, of being a mere collecting of diverse and pretty objects, without any higher philosophical motive. With a certain class of collectors who "covet" specimens it never becomes a rational employment, although, by the constant acquisition and sale of material, it may be made a lucrative one. As indulged in by such persons it loses much of its refining influences and educational value, and becomes merely the opportunity for the display of human passions and idiosyncracies.

Among the different species of Noctuidæ which I have watched from the egg to the moth-stage is *Aletia Argillacea* of Hübner, the "Cotton-Worm" of the Southern States. As this species has interested me for many years, I give my observations upon it from the Alabama Geological Report, and they will apply, generally, to the mode of development in the family Noctuidæ. The cotton-worm is, in its earliest stage, a fertilized egg, which is deposited by the female moth on the leaf of the cotton-plant. Within this egg, which is so small as not to be readily perceived, the growth of the young "worm" rapidly proceeds, until in a few days it is large enough to commence its free existence, and escapes by eating its way through the "shell." If we now examine this worm or larva, we find that the body is made up of successive rings. The first three of these "rings" or segments, behind the head, bear each a pair of horny, jointed legs, six in all, armed with bristles and terminating in a claw. If we compare the cotton-worm, in this stage, with the common rain- or earth-worm, for instance, we see that it differs by possessing these jointed legs, although the bodies of the two animals are alike in being made up of successive rounded rings or segments with a fold between them. They belong, in fact, to two different types of structure; the cotton-worm being an Arthropod or jointed-foot insect, and the rain-worm belonging to the true footless worms or Vermes.

Counting backwards from the head, we find that on the 6th, 7th, 8th, and 9th segments of the body of the cotton-worm there are pairs of short fleshy projections, which are not jointed, but are used by the cotton-worm for progression. The pair on the 6th segment are not used, and there are projections of the skin on the 10th segment also, showing a distribution of these fleshy processes or false legs along the line of the hinder part of the body, without reference to their usefulness to the animal, and in an imperfect condition of development. The last segment of the body is provided with a pair of these fleshy false legs for grasping the leaf and maintaining the position of the animal while feeding. When we come to examine the anterior end or head of the cotton-worm, we find it made up of a harder covering above, and beneath of pairs of jointed appendages, the most prominent of which are the cutting-jaws or mandibles, which perform the office of supplying food by tearing off the leaf of the cotton-plant. These jointed appendages to the head are similar in structure to the jointed feet of the animal, though they serve a different purpose in its economy; they are here head organs; so that we now see that there are three distinct regions of the body in the cotton-worm characterized by three different kinds of appendages. These different regions are technically called head, thorax, and abdomen. In walking, owing to the disuse of certain of the abdominal or fleshy false legs, the cotton-worm doubles the body between the thorax (which bears the true jointed legs) and the 7th abdominal segment. This position of the body gives it the name of a half-looper. As it grows, the yellowish-green cotton-worm casts its skin from time to time, feeding all the while and growing rapidly. The segments of the body are seen to be ornamented with black dots, which, under the microscope, appear as warts, some of which give rise to hair. In some of the worms there is a distinct dorsal line visible, wanting in others. This stripe down the back gives the worms a peculiar appearance: it seems to be a variation, the colour due to the massing of pigment-cells in the skin, and not a reflection of the digestive system of the animal, which begins at the mouth and ends at the last segment, through the opening of which the leaf-food of the animal is expelled in little pellets. This variation of the markings of the cotton-worm is interesting, because it shows the worm to be undergoing some slow process of modification; and it may be that its present mode of life in the Southern States is producing some change in itself.

In Central Alabama, I have watched the growth of the worms on the cotton-plant. The worm appears there in certain seasons as early as the latter part of June. After feeding for a period of about fourteen bays, the cotton-worms begin preparations for shedding their skins to pass into the chrysalis stage of growth. For this they spin a few loose threads of silk on the plant itself, which they rarely forsake for that purpose. Within this light web the last larva-skin is thrown off, and the brown chrysalis-skin is exposed. In this state the worm passes from a week to ten days. During this time, although appearing quiet outwardly, and without exterior organs of locomotion, growth takes place within the shell of the chrysalis. At last it has progressed so far that it arrives at maturity. Through an opening of the head and thorax of the chrysalis the full-grown cotton-fly or moth appears, its wings merely little pads at the sides. These are quickly expanded by a muscular action, and by a circulation in the veins of the wing, which ceases so soon as the wings are dried in the sun. The body is now found to be covered with scales; the wings cover the body so much, that at first we cannot see that it is, after all, the same animal which we knew first as a larva. But the three portions of the body may be seen. The head has two long jointed antennæ or feelers. The jointed maxillæ have become a spiral tongue. The thorax supports its six legs as before, while the fleshy or false legs of the abdomen have disappeared, as being of no further use to the animal. The insect is now mature, and in a condition to commence its work of propagating its young.

The migrations of animals are among the most important circumstances affecting the forms of life. Wallace and Wagner have already shown how the separation, in this way, of local races or varieties may have given rise to new species. There is first to be considered the involuntary migrations of animals by being floated down rivers or conveyed by the wind. As we study those which are provided with wings, their voluntary migrations are seen to play an important part in their life. Birds and insects share these characters in common. Flights of the Storm-Butterfly (*Danaus Plexippus*) have already been noticed crossing the Great Lakes in the autumn, and going southward as if to endeavour to hibernate in a warmer climate. This butterfly hibernates in Alabama. Flights of butterflies have been frequently observed in regions as remote as the English Channel and the Amazon river. The cotton-worm moth is strong-winged and has a lithe, smoothly scaled body, offering very little resistance to the wind. Although the wind may accelerate and assist its migrations, I regard them as voluntary, from the facts of its structure and the wide territory which it covers. The cause of the northward migration of the cotton-worm moth from more southern localities over the cotton belt, and as far north as Canada, cannot be suggested as yet, the data not being all known. One thing is clear, that the territory growing cotton, over which they pass, increases their numbers by providing them with food, and thus makes them an enemy of the cotton-planter.

In the United States the yearly condition of the crops has an immediate effect upon the general prosperity of the country. Any thing affecting agricultural interests thus becomes a matter of public anxiety, and it is so with regard to the injuries inflicted by certain insects upon the crops. It is now twenty-two or three years ago since my old friend Mr. Townend Glover, then Entomologist to the Department of Agriculture, sent me specimens of certain Noctuidæ to determine, which had been collected as injurious to different crops. Among them were specimens of the cotton-worm, which had been described by Thomas Say under the name *Noctua Xylina.* At that time there were very few Noctuidæ named in any of the Museums in the country; and of the nearly fifteen hundred species of Noctuidæ now mentioned in our books, not fifteen were known by name in any collection in the United States. The descriptions in French of M. Guénée had not been translated; a few species had been described by Dr. Harris in his Report on the Insects injurious to vegetation made to the Massachusetts Legislature; but in no public or private collection in the United States were there more than a dozen kinds of our Noctuidæ properly named. And it was certainly difficult to obtain any information as to what had been done by European writers in the group. At the present time the general knowledge has increased, so that from 600 to 800 species from the East are well known to entomologists, and the immature stages of a considerable number have been discovered. To a large extent the publications by the Department of Agriculture and the different States have assisted in bringing about this improved condition of affairs. The treatise of Dr. Harris, which has become classical on its subject, did much towards creating a general interest in Entomology. But the publication of the 'Canadian Entomologist,' a journal aided pecuniarily by the Dominion Government, and owing its success chiefly to the unselfish labours of Mr. William Saunders, has assisted the progress of Entomology in America probably more than any one other similar undertaking. The publications of the Entomological Society of Philadelphia must, however, not be overlooked; their success was owing to the care and attention of Mr. Ezra T. Cresson, their establishment largely to the generosity of the late Dr. Thos. B. Wilson, whose name is recalled in our collections by the pretty and interesting moth *Ciris Wilsonii.* But the real fostering influence of Entomology in America is its practical side, the interests of agriculture, although the condition of political machinery in the United States has allowed a different set of qualities than purely scientific ones to influence its appointing power. The first journal published in the United States devoted to Economic Entomology was, I believe, the 'Practical Entomologist,' issued by Mr. Cresson, Mr. Blake, and, perhaps, other gentlemen connected with the Entomological Society of Philadelphia, and of which I was Editor for the first few numbers *, to be succeeded by the late Mr. B. D. Walsh.

The actual amount of injury inflicted by insects on the yearly crops of various kinds in the United

* Among the editorials contributed by myself was a short one entitled "*Cui Bono?*"

States is, as may be imagined, from their immense extent, very great, aside from such occasional severe visitations as that of the grasshoppers in the West. Yet it is difficult to give any exact calculation in figures of the amount of damage thus inflicted. Take the case of the Cotton-Worm, with which I became familiar from observing it for several years upon my own and my mother's plantations in Central Alabama, and which, being one of the Noctuidæ, is pertinent to the subject of this Essay. This insect spreads during the season, from south to north, over the cotton-growing region from Texas to the Ohio River. The larva strips the plant of its foliage, and, where it occurs in force early in the season, and remains in the locality through successive broods, it inflicts great damage, eating finally the flowers, the soft bolls, and cutting off the last picking of cotton. Yet its appearance is not uniform over the region; it rarely greatly increases in number until after the main crop is formed, and where it attacks cotton on bottom lands, growing rank and large, it does but little real harm. By eating the leaves it causes the later bolls to mature more rapidly; while in South-western Georgia and some parts of Alabama, the "rust" (a vegetable parasite, of which we hear comparatively little) is a much worse foe of the cotton-planter than the "worm." It will be readily seen how the injuries committed by any one insect can be over-estimated upon paper by interested parties, when we see how much should enter into the calculation. The history of the cotton-worm is the subject of an extensive volume by Professor Comstock, recently published by the Department of Agriculture, and a shorter one by Professor C. V. Riley. Its full history is not yet ascertained, the territory which it covers in its migrations being very extensive. After having studied the worm in the South since 1868, and having brought the results before the public in a lecture, I endeavoured, in 1874 and subsequently, to interest the Government in the work of collecting all possible information on the subject, and advised the gradual accumulation of facts relating to the time of appearing, &c., by means of the Postal and Weather Service. Much, I am satisfied, remains to be done in studying the relation of its appearance to the prevailing winds. It is probable that repressive measures taken early in the season at certain points to be ascertained would sensibly mitigate its later ravages. The results of the two different inquiries, undertaken under the management of other parties, are before the public; but the statement of Professor Riley as to what has been really done towards the protection of the planter seems to me as exaggerated as his various publications on the subject are premature and unfair to other scientists. The labours of Mr. Townend Glover, for instance, his discovery of the attraction presented by the glands of the cotton-plant to the moth, &c., have been appropriated; and there has been too much capital manufactured by Professor Riley at the expense of others, for the mere purpose of making his own position secure. After having ordered Paris Green for almost every injurious insect, and claiming public gratitude for his procedure, it is probable that the Government will have to be called upon to interfere in the matter of the reckless use of this arsenical poison. Already much valuable stock has been sacrificed to the preservation of the potato-plants, which a little care and industry would have otherwise protected; and cases of the poisoning of farmers and their families by maliciously disposed "hands" and servants have been reported in the daily press. For some years I have been calling public attention to the reckless use of Paris Green as an insecticide, especially fearing its introduction into the South. It is true that the use of Paris Green has saved a good many crops of potatoes in the United States from the attacks of the *Doryphora* or Potato-Beetle; but lime applied to the young insects, and an industrious use of the beating process, would have effected the same result. When applied to the cotton-plant, as Professor Riley recommends, Paris Green is open to objections. Under their good nature and general acquiescence in their condition of life, to which their extraordinary adaptiveness and capacity for copying the manners of the whites assists, the negroes in the Southern States have shown a certain readiness for the commission of revengeful crimes; and the wholesale use of such a poison as Paris Green on plantations would give them a ready and suggestive instrument to their hands. Being a mineral poison, and filtering into the soil, Paris Green is also dangerous in a country where surface-wells are used.

The extent of the present use of Paris Green and arsenical preparations in the United States is not generally known in the absence of proper statistics. Were these published, public attention would probably be aroused to the danger of the situation.

With every other field for human inquiry and action, the study of even such a small assemblage of insects as the Noctuidæ touches a number of other interests which at first sight appear remote. But the Noctuidæ of North America include at least two species of insects injurious to agriculture which, in this respect, are among the most important of insect enemies to man on our continent. Beside *Heliophila Unipuncta*, the Army Worm, and *Aletia Argillacea*, there are a certain number of species of cutworms belonging to the genera *Agrotis* and *Hadena* which attack various field- and garden-crops. Outside of these are the great bulk of the species of Noctuidæ which feed on weeds and plants not of any particular economic importance. There are plenty of them to reward the labours of the collector, and to puzzle the philosophers who believe that every thing has its use, and that man himself is the pivot about which all creation turns.

I cannot close this Preface without thanking a number of correspondents who have sent me material during the last twenty-five years from various points of the United States and Canada. I owe almost all my knowledge of Western Noctuidæ to the collections received from Mr. Henry Edwards, Mr. James Behrens, Prof. F. H. Snow, Mr. Theo. L. Mead, and Mr. Berthold Neumœgen.

In the East Mr. Roland Thaxter, Mr. L. W. Goodell, Mr. G. R. Pilate, Prof. C. H. Fernald, Mrs. Fernald, Mr. Fish, and Mr. Allen have been very kind to me in communicating material. Mr. Coquillet has sent me notes on many larvæ. Prof. J. A. Lintner, Dr. James S. Bailey, Mr. Hill, Mr. Gray, and Mr. von Meske have sent me many species from the central portion of New York State, as well as from the North Woods. From Rhode Island the collections of my kind friend of long standing, Mrs. S. W. Bridgham, have allowed me to examine most of the species occurring near the seaboard of the Eastern States. No collection that I have seen has been brought together with more care; and by the gathering of many specimens of a species, Mrs. Bridgham has given me important data for studying variation in this group. From the South I have seen most of Belfrage's collections in Texas, and those of Mr. Schwarz and Mr. Koebele in Florida. I have collected myself in Alabama, and Mr. Morrison has sent me a quantity of Deltoids from North Carolina. I have acquired some of Mr. Morrison's collections in Southern California and Washington Territory; and Mr. Fred. Tepper and Mr. E. L. Graef have kindly allowed me to examine the species they have received from the same collector.

It was a labour of love to Mr. Henry S. Sprague, the Entomologist, when he drew the excellent Plates of Noctuidæ, on stone, which illustrate my articles published by the Buffalo Society of Natural Sciences.

Mr. Julius Pohlman, my assistant during the

concluding portion of the seven years in which I held the position of Director of the Museum in the Buffalo Society of Natural Sciences, has made several beautiful drawings for me, and was a kind and intelligent aid in my entomological work. I cannot forget my departed friend Hochstein, unrivalled as a painter of butterflies and moths.

In this brief *résumé* of the sources from which I have drawn information, I have no doubt omitted to mention the names of several entomologists who have corresponded with me. As I write I recall Mr. James Angus, who sent me some fine Catocalas from his captures at West Farms, N.Y., and Mr. Dury, who has collected many species at Cincinnati. There is no more pleasant task than that of recalling all the friends one makes in connection with a pursuit like Entomology; but with my acknowledgments to Mr. George Norman, of Cluny Hill, who spent some time in Canada successfully collecting Noctuidæ, my list must end. Besides discovering *Lithophane Georgii*, two pretty species, *Agrotis Normaniana* and *Crocigrapha Normani*, were among Mr. Norman's captures, and will serve to remind him of American Entomology.

STRUCTURE AND LITERATURE.

THE moths belonging to the Family called Noctuidæ, or Noctuæ, by authors, are among the most interesting of the Lepidoptera. Comparatively uniform in shape, and more often of various shades of brown in colour, there are a surprising number of different kinds, separable, usually, by modifications of the pattern of the upper surface of the front pair of wings. As might be expected from the greater extent of territory, the species are more numerous in North America, north of Mexico and the West Indies, than in Europe. I have entered, up to the present time, the names of 1460 species in my notes; while, in a few instances, these names may refer to the same species, it is evident, from the partial exploration, that a great many remain to be discovered. I think, finally, that about two thousand species of Noctuidæ will be found in our territory.

There is comparatively more agreement among authors as to the limit of the Family than in some other cases in the same Order of Insects; but the differences are still great, arising from the different characters regarded as essential in deciding the matter. M. Guénée included a small group, the *Bombyciæ* of Hübner and *Cymatophorina* of Herrich-Schäffer, as well as the genus *Brephos* of Hübner, in the Family, while excluding the Deltoids. Dr. Herrich-Schäffer, who was the first author to use the veining of the wings as giving decided family characters in the Lepidoptera, excluded the *Cymatophorina* and *Brephina*, while including the Deltoids. The late Julius Lederer followed Dr. Herrich-Schäffer in this course. He regarded as essential characteristics of the neuration of the Noctuidæ, that vein 5 of the primary wing should be nearer to 4 than to 6; and that there should be two internal veins on the secondaries, which have besides seven other veins: 6 and 7, arising from the upper and outer corner of the discal cell, and 8 from the base of the wing, soldering more or less plainly with the subcostal vein at the base.

Dr. A. S. Packard, jun., considers the characters taken from the appendages as indecisive in establishing the families of moths, which he regards from the point of view in which they were established by Latreille. Dr. Packard depends upon the relative size and shape of the clypeus, or front, between the eyes, as the best distinguishing mark. This he finds in the Noctuidæ to be "about as long as broad, narrowing slightly towards the front, where it is emarginate; the anterior edge is often turned up; surface full, convex, smooth, the convexity greatest just below the middle, sometimes becoming a tuberosity" (Proc. P. S. N. H. vol. i.). Dr. Packard apparently follows M. Guénée in his limitation of the Family. In my own lists and papers on the Noctuidæ of North America, I have included the Deltoids with them, not being able to find any distinguishing characters; and I have regarded the *Cymatophorina* and *Brephina* as subordinate groups of the Noctuidæ, not being satisfied that the neuration gives a determinative family character. In *Nolaphana*, which seems otherwise a Noctuid, vein 5 of the fore wings is situated midway between 4 and 6.

When we turn to the arrangement of the Family into subordinate groups, or subfamilies, we find that opinions again vary. Generally speaking, authors seem to have recognized two divisions—M. Guénée calls them *Trifidæ* and *Quadrifidæ*, and Dr. Packard *Noctuinæ* and *Catocalinæ*. The earliest designation of these groups appears to be by Borkhausen, who calls them *Nonfasciatæ* and *Fasciatæ*, from the differing pattern of ornamentation; and these terms I have adopted myself. Lederer, however, will admit no subordinate groups whatever. Having thrown out *Cymatophora* (*Bombycia*) and allies, together with *Brephos*, he allows the genera to follow one after another, including the Deltoid forms, without a break from *Diloba* to *Rivula*.

It seems to me that the Noctuidæ are a large group of but slightly differing structural forms, lying between the Geometridæ and the group Notodontidæ, or *Ptilodontes* of the Bombycidæ, and with a more remote connection with the Sphingidæ. Mr. A. G. Butler says that "the Noctuidæ are in all probability descended from the Geometrites; that their ancestors were 'loopers.'" I find that there is a constant modification of the markings and form in the Noctuid moths, in the two following directions. On the one hand, the upper surfaces of the hind wings tend to resemble the lower surfaces of the primaries: they are dull, unbanded except by one or two exterior cloudy lines. In position these moths hold the primaries over the secondaries, shielding the upper surface of the latter from the light and air during the daytime. It is as if the pattern of the one was photographed upon the other. In the other direction, the ornamentation tends to be uniform on the upper and under surfaces of both pair of wings. The transverse lines run across both wings, so that the upper surfaces of the primaries and secondaries come to look alike. In this case the moths rest in the daytime with the hind wings more or less exposed and after the fashion of the Geometridæ. In comparing the method of variation, I have found that in the Noctuidæ, especially those belonging to the first group, or *Nonfasciatæ*, representative forms differ first, and most strongly, in the appearance of the upper surface of the primaries, then in the secondaries, and lastly in the under surface of the wings. Thus the American *Catocala Relicta*, which belongs to one of the higher genera among the *Fasciatæ*, approaching the *Nonfasciatæ* in the position of the wings at rest, and which represents the European *C. Fraxini*, differs more or less noticeably by the fore wing above. On the hind wing the dusky blue band of the European species becomes white in the American; while I have previously shown that, in some specimens, there is a narrow blue edging retained in *C. Relicta*, throwing light on the origin of the species (Can. Ent. viii. 231). In the separation of the forms, exposure to a changing environment has assisted in giving character to the species. For the results of some very careful studies on the characters of allied species in the North-American and European Noctuidæ, the papers of Dr. Speyer in the 'Stettiner ent. Zeitung' should be consulted.

Many of the North-American species more or less closely resemble European insects. There is an almost perfect gradation between absolutely undistinguishable forms, occurring on both continents, such as *Xanthia Silago*, to perfectly dissimilar ones. Again, the caterpillars seem to have submitted to independent modification, while the moths produced by them remain comparatively unaltered, *e. g.* the genus *Acronycta* (*cf.* Ann. N. Y. Lyceum N. H. Vol. xi., Article xxviii). All these facts, and others presented by myself in the 'Bulletin' of the Buffalo Society of Natural Sciences, point to the arising of species by derivation. When we turn to the question of the distribution of the forms of Noctuidæ, we must look to former geological epochs for most of the explanation. The North-American Noctuidæ are evidently descended in great part from a former circumpolar fauna during the Tertiary Period. In a paper read before the American Association for the Advancement of Science in 1875, and printed in Silliman's Journal, 3rd series, Vol. x., No. 59, I brought together facts to show the way in which the Glacial Period has influenced the present distribution of our North-American insects. I also reprint, with this Essay, a more popularly written paper, entitled

"A Colony of Butterflies," which contains the results of my studies on this subject, and which first appeared in the pages of the 'American Naturalist.' For myself I do not believe that there is any basis of fact behind the myth of the Atlantis. I do not think that there is any probability of a former continent between Europe and America, a bridge for the fauna or flora and at the same time a surviving memory in men's minds. Rather does it seem reasonable that the observation of low-lying clouds, in a sun-flushed, western sky, suggested the fabled countries. The birth-place of the myth seems to have been far from the shores of the Atlantic. The myth must be separated, at all events, from any geological evidence of the former existence of an Atlantic continent.

At this time a large portion of the western and south-western portions of the United States remains unexplored, so that we have much to learn as to the geographical distribution of our Noctuidæ. A good number of species are found to range from Texas, through Arizona, and into California. There is a general distinctive character to the Noctuid fauna of the Eastern and Middle States as far as the Mississippi River; and, on the whole, the Eastern or Atlantic fauna resembles the European less than the Western or Pacific. There is also a seasonal migration, from the South to the North, of many species. The topography of the land is consulted in their flights; and the Southern species reach their highest northern extension along the coast or up the valley of the Mississippi River. *Euthisanotia Timais*, the Spanish Moth, which (as found by Mr. Thaxter) breeds in Florida, is found on the coast of Long Island; *Erebus Odora* is found sometimes in Maine; on the other hand, *Thysania Zenobia* and *Brotis Vulneraria* have been collected in Iowa and Wisconsin. These are immigrants, and probably do not breed within our territory. On the other hand, some species seem to occur from Guiana to New York, such as *Homopyralis Tactus.* The fauna of the southern extremity of the peninsula of Florida is much like that of Cuba and even Jamaica. The genus *Catocala* is more largely represented in the United States than elsewhere; as remarked in my paper on that genus, it seems to be confined to the Northern Hemisphere, and does not cross the Equator. As we progress southward the lower forms of the Family, *i. e.* the *Fasciatæ*, or *Catocalinæ* of Dr. Packard, become more numerous in genera and species.

Our knowledge of the Noctuidæ of Texas and the South-west and West has been increased by the collections of the late Mr. Jacob Boll in Texas, as well as those of Mr. Belfrage. Mr. Theo. L. Mead has made some fine collections in Colorado, discovering the yellow-winged American forms of *Oncocnemis* described by myself. Mr. Berthold Neumœgen has collected in the West, and employs a naturalist in Arizona, whence many rarities have come, described in part by Mr. Henry Edwards, who has himself made extended collections on the Pacific Coast. Prof. F. H. Snow, of the University of Kansas, has made some very fine collections in Colorado and

New Mexico, in the latter region not without risk of his life from the Apache Indians.

In classifying the Noctuidæ the following characters have to be observed. The structure of the antennæ, which, in the males, are often pectinated, or have brush-like surfaces beneath. The presence or absence of ocelli, situate behind the antennæ near the margin of the compound eyes, must be ascertained. The surface of the compound eyes is either studded with hair or naked; the rim of the eye is fringed sometimes with longer hair-like scales. The clypeus between the eyes is flat or bulging; again, it is horned or furnished with a pit or depression. The tongue is usually long, but sometimes weak and short. The labial palpi are subject to some variation in the comparative length of the joints and their vestiture. The tibiæ are either spinose, furnished with prickles, or unarmed; the front tibiæ are sometimes swollen, or, again, very short; the spurs on the middle and hind tibiæ must not be mistaken for these spines or thorns. In addition the front tibiæ exhibit a varied structure: they have a simple claw-like spine at the extremity (*Oncocnemis* and the Dicopid genera), or there are additional smaller ones, as in the Heliothid group. Again, the joint is terminated by a tridentate extension of the tegument, as in *Triocnemis*, to speak alliteratively. The vestiture of the body should be examined under the microscope and its nature ascertained. In *Bryophila* and *Chytonix*, as also in *Tarache* and *Tamila*, the scales are in part or wholly flattened, not hair-like. Then the crestings of the thorax and abdomen on the dorsal line need attention, and for this *perfect* specimens are requisite. The wings may have the outer margin scalloped (*Trigonophora*), even (*Conservula*), uneven (*Brotolomia*), or angulated (*Scoliopteryx*). There is a variation in the costal curve, and the apices may be either sharp, pointed, or produced. As a rule the wings are short rather than long; but in *Cleophana* and *Cucullia*, especially the latter, the wings are long and narrow. In approaching the lower group, or *Fasciatæ*, the wings broaden. The veining is usually but little varied. The fore wings have twelve veins, counted on the external and costal margins, and usually a small accessory cell above and beyond the discal cell. Variations are offered by the method of branching of the subcostal nervules; and in a paper published in the 'Canadian Entomologist,' vol. xi. p. 231 *et seq.*, I have discussed these variations in closely allied genera; they chiefly affect veins 7 to 9. In a few genera the males have an aberrant neuration, and this becomes a sexual character. Such, for instance, is the genus *Heliochilus*, found from Alabama to Colorado, and figured by me in the 'Proceedings of the Entomological Society of Philadelphia,' vol. iv. plate ii. figs. 3 ♂, 4–5 ♀. The subcostal vein is bent downwards, leaving a large space above the discal cell and between it and the costa, which latter is thickened about the centre (see figure of venation, *ibid.* p. 328). The Southern genus *Pteraetholix*, collected by me in Alabama, is also characterized by a large pellucid impression near the base of the discal cell in the males; underneath, the edges, especially outwardly, are overlapped by the tegument. The median vein is bent downwardly in a curve at base to make room for the impression, and thus widens the discal cell. *Pteraetholix Bullula* is a mixed reddish species, smaller than *Heliochilus Paradoxus*, which, in its colour and markings, is much like the larger *Heliothis Umbrosus*, under which latter name it appears I have described American specimens of the European *H. Armiger*. In another form, *Chytoriza Tecta*, Grote, Can. Ent. viii. p. 190, there is a small pellucid impressed spot on the male primary wing. This genus, in which the wings are a little wider, seems related to *Pteraetholix*, *Anomis*, and *Aletia*, while *Heliochilus* is allied to *Heliothis* and *Melicleptria*. While in the other genera the wings are kept stiff when at rest, in the curious genus *Marasmalus*, with two species, *Ventilator* and *Histrio*, found from Massachusetts to Texas, the wings are rolled together so as to conceal their true size. When these insects alight, they seem to disappear upon folding their wings; at such times they may be fancied to "have the receipt of fern-seed" and to "walk invisible." The genus is related to *Pencillaria* and *Eutelia*. Of this latter we have a beautiful and rare species in the United States, *E. Pulcherrima*, found in New York and New Jersey hitherto.

The discal cell on both wings in the Noctuidæ is open or partially closed by a thickening of the tegument. On the hind wings vein 5 is sometimes free and again connected by this cross vein with the series of median nervules. It is variably strong and sometimes wanting; I have shown how it varies in the genera related to *Erotyla* and *Spragueia*. It has no value in establishing the subfamilies, and its variability suggests that, except in a general way and in relation to the shape and form of the wings, the particular course and appearance of the veins cannot be used to establish higher groups than genera; in these latter the characters offered by the veins are often convenient to use. An examination of the shape of the outer corneous pieces protecting the male genitalia shows that occasionally they offer peculiar shapes in different genera. The pattern of the wings is sometimes peculiar, and assists in the work of locating the species; but it is, on the whole, very uniform. The colours vary somewhat and, in the genera allied to *Heliothis*, which frequent flowers in the daytime, they are often very gay.

With regard to the immature stages, the North-American Noctuidæ offer several striking larval forms. The caterpillar of *Harrisimemna Trisignata* is remarkable for its resemblance to that of the Diurnal genus *Basilarchia* of Mr. Scudder. The larva has long, somewhat isolated thoracic hairs, elevates the front of the body when at rest and, when disturbed, sways the free portion from side to side, which has caused it to be known as the "zig-zag caterpillar." It lives on species of *Syringa*, and pupates by boring into the solid wood (in confinement a piece of solid pine will serve its purpose), carefully closing the orifice behind it. Professor J. Henry Comstock ('Papilio,' i. p. 147), who has discovered a true predaceous larva among the Phycidæ (N. Am. Ent. i. p. 25, with Plate), has also found an aquatic larva among the Noctuidæ. It is that of *Arzama Melanopyga*, one of a genus belonging to the Nonagrians, a stout Bombyciform moth having the abdomen of the female provided with a large, terminal, close tuft. The larva is furnished with nine pairs of spiracles, lives in the stems of pond-lilies, and can remain below the surface of the water a long time.

As to mimicry, the Noctuidæ which are grey or brown often look like lichens or bark when resting on the trunks of trees in the daytime. A curious example of protective mimicry is given by Professor Kellicott in the pages of the 'North-American Entomologist,' i. p. 30 (Oct. 1879). It is that of the pretty Noctuid *Rhodophora Florida*, which lives concealed in the daytime in the withering blossoms of the evening primrose (*Œnothera Biennis*). The moth has the inner two thirds of the fore wings bright pink, while the outer third, the hind wings, and abdomen are pale yellow. It enters the flowers before day with its body resting upon the *style*, the four-parted *stigma* projecting beyond the top of the abdomen, appearing like a part of it. When the sun comes, the two petals that were above the moth soon wilt, and fall down over the roof-like wings concealing the hind portion, leaving the yellow part exposed as part of the blossom. Sometimes the pink of the wings is not wholly covered, but the tone of the continuous colours is such that the harmony is complete. The larva, which feeds on the floral organs and seed-pods of the same plant, resembles the flower-spike in its yellowish-green colour; it is also clothed with short white hairs, making the surface pubescent like the plant. Prof. Kellicott's observations on this species could doubtless be repeated with many others, and are very opportune. The grey larvæ of *Catocala* and the grey front wings of the moth conceal themselves readily against the trunks of the trees, or the branches where they usually rest. The colours of the American species of *Glœa* and *Scopelosoma* are brighter, as compared with the European, and thus follow the example set by the autumn leaves in which these moths conceal themselves by day.

In May 1874 I published a List of the North-American Noctuidæ, classifying them as nearly as I could according to Lederer's method, which I had applied to many of the species in a number of separate papers in various scientific journals. In 1875–76 I published a Check List of the species, and have since then been working on a fresh Catalogue, giving full references and adding what is known as to larvæ, food-plants, and locality. I have published lists of detached genera in advance of the Catalogue, which I may not be able to complete. In the pages of the Bulletin of the United States Geological Survey, I have catalogued the genera *Hadena*, *Agrotis*, and *Polia*; in 'Papilio' I have catalogued *Eustrotia* and *Oncocnemis*; in the 'Canadian Entomologist' the genera *Mamestra*, *Orthosia*, and genera allied to *Lithophane*, *Tarache*, and several others. Mr. Henry Edwards has in preparation a list of *Catocalæ*.

While there are a good many questions of synonymy to settle, among which Mr. Walker's species are not the least of the difficulties, yet a great deal of preparatory work for the new Catalogue is already accomplished. The student will find most of the published information respecting the North-American Noctuidæ, issued during the last twenty years in America, in the pages of the following journals and publications:—

The Canadian Entomologist.

The North-American Entomologist.

The Proceedings of the Entomological Society of Philadelphia.

The Transactions of the American Entomological Society.

The Proceedings of the Boston Society of Natural History.

The Bulletin of the Buffalo Society of Natural Sciences.

Reports of the Peabody Academy of Sciences, Salem, Mass.

The Proceedings of the Academy of Natural Sciences of Philadelphia.

The Transactions of the Kansas Academy of Science.

The Bulletin of the U. S. Geol. and Geographical Survey of the Territories.

Reports on the N. Y. State Cabinet of Natural History.

Check List of the North-American *Bombyciæ* and *Noctuelitæ*.

Bulletin of the Brooklyn Entomological Society.

'Psyche.'

'Papilio.'

NOTES ON MR. WALKER'S TYPES OF NORTH-AMERICAN NOCTUIDÆ IN THE BRITISH MUSEUM.

By the favour of Dr. Albert Günther I have been permitted to view, for the second time, Mr. Walker's types of North-American Noctuidæ in the British-Museum Collection. I have not been able to examine the species here mentioned with the microscope, but I have given my opinion on what was clearly to be ascertained from a mere viewing of the specimens, which, incomplete as it is, cannot fail to be of interest to the student in America, from the vague and unsatisfactory character of Mr. Walker's descriptions.

Acronycta Cristifera, *Walk.*

The type, from Hudson's Bay, is not an *Acronycta.* The abdomen is tufted; the species is dark stone-grey, with kidney-shaped reniform, and seems a Hadenoid form unknown to me.

Acronycta Fasciata, *Walk.*

Four specimens under this name are *A. Brumosa* of Guénée (*Verrillii* m.).

Acronycta Impressa, *Walk.*

The type, from Hudson's Bay, seems to be a small specimen of *Brumosa* ; it is rubbed, and apparently differs a little from that species ; it is hardly recognizable.

Acronycta Contacta, *Walk.*

Two specimens, from Hudson's Bay, belong to the species described as *Acronycta Aspera* by Mr. Morrison, and *Polia Diffusilis* by Dr. Harvey. It is not an *Acronycta*. The species seems a Northern one, and has been taken by Mr. Hill of Albany in the Adirondack region of N. Y. State. It may be known in future as *Polia Contacta*.

Acronycta Grisea, *Walk.*

I have hitherto correctly identified this species in my own and other collections.

Acronycta Mixta, *Walk.*

The specimen, from St. Martin's Falls, is *Agrotis Speciosa* (=*Polia Perquiritata* of Mr. Morrison).

Acronycta Circulifera, *Walk.*

As formerly stated by me, the specimens belong to *Charadra Deridens*, Guén. Mr. Walker had a specimen of *Jocosa* wrongly named "*Deridens*."

Acronycta Spinigera, *Guén.*

Two (the larger specimens) belong to *Morula*; the middle one is *Harveyana*. Generally speaking, Mr. Walker's identifications of Guénée's species do not accord with my own.

Bryophila Discitincta, *Walk.*

I do not recognize this species. It is not a *Bryophila.* It is a dark, bluish-grey form, and may be an *Agrotis.*

Bryophila Discivaria, *Walk.*

Two specimens, from St. Martin's Falls, belong to *Parastichtis,* and are respectively light and dark varieties of *Gentilis.*

Bryophila Discinigra, *Walk.*

I do not recognize this species. It is decidedly not a *Bryophila.* It is a rather stout form, much shaded with black, reminding me a little of *Valeria* ? *Conserta,* though different.

Cymatophora Viridescens, *Walk.*

The specimen from Florida is a *Dicopis,* the fore wings quite greenish, and close to *D. Muralis,* while smaller.

Mythimna Obusta, *Guén.*

Evidently Guénée's type. A *Heliophila,* and, apparently, a red variety of *Pseudargyria,* but darker than any I have seen.

Mythimna Ebriosa, *Guén.*

Evidently Guénée's type, and unknown to me. Bright purple-red, with a pale mark for the renifrom.

Mythimna? Littera, *Guén.*

The specimen is evidently Guénée's type, and, as I have long suspected, is my *Pseudolimacodes Niveicostatus*, an insect which is certainly not a *Mythimna*, and must be known in future as *Pseudolimacodes Littera*. It belongs, most probably, to the *Fasciatæ*.

Mythimna Decolor, *Walk.*

A discoloured specimen, apparently belonging to *Cosmia Infumata*.

Mythimna Subporphyria, *Walk.*

The specimen is broken, and not, I think, a *Mythimna*. It may be found to be an *Agrotis*, when the tibiæ are examined. I do not recognize the species. There are no markings; the median lines vague, broad, darker shades.

Mythimna Contraria, *Walk.*

This is *Mamestra Picta*, Harris.

Mythimna Tripars, *Walk.*

Not a *Mythimna*, and probably an *Agrotis*. Not known to me.

Mythimna Vetusta, *Walk.*

The specimen is a much rubbed *Agrotis*; one of the white species, perhaps *Muræenula*. From these identifications the student will recognize the fact, that it is simply impossible to use Mr. Walker's work without the specimens at hand to find out what is meant by his Latin names; while, from the state in which they are in, one is often left not much wiser than before.

Under the names *Extincta*, *Linita*, and *Insueta*, different species are mixed up. Some of them may really be Guénée's species; but the work of that Author must be compared.

Leucania Diffusa, *Walk.*

The specimen is one of *Harveyi* (*Albilinea* of Guénée; I do not feel sure that it is Hübner's), in very bad condition. The specimen determined as *Videns* (Guénée's type?) is not a *Heliophila*, but *Platysenta Atriciliata*; and *Nonagria*? *Indigens*, without locality, is the same species.

Eudryas Stæ. Johannis, *Walk.*

This is evidently a good species; the hind wings have no terminal band, the fore wings are darker at base than in *Grata*. The species is in no North-American collection I have seen, and the locality is doubtful.

Hydrœcia Salicarum, *Walk.*

This is *Agrotis* (*Pachnobia*) *Orilliana*. The question as to whether my name or Mr. Morrison's should obtain for the species is settled by Mr. Walker taking it. It should be known for the future as *Agrotis* (*Pachnobia*) *Salicarum*. Mr. Walker quotes Dr. Barnston's MSS.

Edema? Obliqua, *Walk.*

This is *Sphida Obliquata*, G. & R.

Mamestra Contenta, *Walk.*

This is *Hadena Devastatrix*, Brace.

Mamestra Ordinaria, *Walk.*

This is *Hadena Devastatrix*, Brace.

Mamestra Unicolor, *Walk.*

This is *Agrotis Clandestina*, Harris.

Mamestra Insulsa, *Walk.*

The specimen, from Canada, is evidently an *Agrotis*, allied to *Repentis*, and unknown to me.

Condica Palpalis, *Walk.*

This species, from Jamaica, is *Hadena Confederata*, which ranges up the coast to New York as a visitant. I am now inclined to place the species in *Perigea*, a closely allied genus to *Hadena*. The species varies greatly, and, as will be seen, is Guénée's *P. Infelix*, according to the B. Mus. Collection.

Mamestra Vetusta, *Walk.*

The specimen is a rubbed, grey *Agrotis*, too poor, I think, to make out specifically.

Mamestra Declarata, *Walk.*

The specimens are Agrotids, allied to *A. Campestris.*

Mamestra Binotata, *Walk.*

The specimen belongs to *Hadena Curvata.* It is rubbed and very dark, with the pale brown reniform contrasting.

Mamestra Plagiata, *Walk.*

The specimen is *Agrotis Bicarnea,* Guénée. It is, I know, almost incredible, but nevertheless a fact.

Xylophasia Indocilis, *Walk.*

The specimens are *Hadena Remissa,* Hübn.

Xylophasia Libera, *Walk.*

The top specimen is *Mamestra Grandis,* Boisd. A second, perhaps a later interpolation, is *Hadena Finitima,* Guén.

Apamea? Insignata, *Walk.*

The specimens are *Hadena Sputatrix,* but Walker's name is preoccupied by himself; and, as elsewhere stated by me, the species must be known by the later name (see Bull. Buff. Soc. N. Sciences, i. p. 190) proposed in consequence.

Apamea Demissa, *Walk.*

The specimen is *Mamestra Latex* of Guénée.

Miana Atomaria, *Walk.*

The specimens are *Telesilla Cinereola* of Guénée.

Miana Undulifera, *Walk.*

The specimen is *Eustrotia Apicosa,* Haw. (=*Nigritula,* Guén.).

Miana Vincta, *Walk.*

The specimen is *Oligia Chalcedonia* of Hübner and my collection; not the var. *Tracta,* which has pale tegulæ and base to the primaries above. It is difficult to understand how these three forms, all previously described, could be brought into one genus not related to any of them, except the last species. However, in America, I have recently heard of a genus "*Ænigma*," with a "large number of costal nervules," and a "*Cosmia*" from Florida, which I believe is *Hyparpax Aurora!* And then there is that old mistake of my own, of which I have lately heard so much that I have really lost all interest in it, the description of a black Arctian (*Geometrica*) as a Zygænid under a new genus. Dr. Boisduval's companionship in a similar case consoled me long ago.

Apamea Remissa, *Walk.*

The specimen is *Perigea Luxa*, Grote; Walker's name is used in the group and for a species in a closely allied genus, perhaps not separable.

The single specimens representing *Celæna Punctifera*, *Infecta*, and *Egens* are really so poor that I do not think the species can be made out with satisfactory certainty.

Celæna? Irresoluta, *Walk.*

This is apparently a bad specimen of *O. Chalcidonia*.

Perigea Infelix.

The specimen (Guénée's type?) is a dark variety of *P. Confederata*, mentioned above.

Perigea Mobilis, *Walk.*

This is *P. Xanthioides*, Guén.

Perigea Paupera, *Otiosa*, *Indicisa*, and *Centralis*

need careful comparison with our *Fabrefacta*, although not described from North America. I think the first four are probably all one species, and the same as *Fabrefacta*.

Caradrina Multifera, *Walk.*

This species is, as I have already stated, the *Fidicularia* of Mr. Morrison. There are a number of good specimens.

I close my Notes here for the present. In the drawers of the first Case, here gone over by me, are several forms of *Nonagria* and *Agrotis*, &c., mostly not in good condition, and which need a careful study of the structural characters of the species in order to verify the reasonableness of any suggestions as to what they are. Some names may be rescued at great trouble; others, I am sure, must be dropped for good. It will be only after much labour that Mr. Walker's names will cease to annoy the student. I have shown here the necessity which exists that this work should be undertaken by a competent hand. It will need a great deal more patience than the description of new species. In order that some conclusion be arrived at, which will allow the study of American Moths to go on with safety to the describers of species, I hope that the Authorities of the British Museum will undertake this work, which, until it *is* done, will always be expected of the Institution which sanctioned the printing of Mr. Walker's Catalogues. It will not be forgotten that through Lord Walsingham's accurate account of the *Tortricidæ*, all the doubts raised by Walker's work in that group are solved. Similar work (it is to be hoped as well conceived) is needed in the other families, and must in time be accomplished.

SPECIMENS OF NORTH-AMERICAN NOCTUIDÆ.

1. Bombycia Semicircularis.

THIS species has been collected in Washington Territory by Mr. H. K. Morrison. The pale, creamy tegulæ and the basal patch, confined to the middle of the primary wing, are distinctive. It differs from *B. Improvisa*, by its quieter colours, the evenly rounded anterior band, and the more continuous apical streak. It is quite distinct from the species from Japan, described by Mr. Butler, in which the transverse lines are perpendicular. As yet this genus in North America seems to be confined to the West Coast, and thus furnishes an example of the greater resemblance of the Western Lepidopterous fauna with that of the Old World. In the Bulletin of the Buffalo Society of Natural Sciences, i. p. 276, is a paper by Dr. Leon F. Harvey on this group of Moths.

Bombycia Semicircularis, Grote, 'Papilio,' i. p. 75.

2. Thyatira Lorata.

This beautiful species is about as large as our Eastern *T. Pudens*. The fore wings are a little broader, and the black, propinquitous, median lines are visible. The apical patch of *T. Pudens* is wanting in *T. Lorata,* in which the basal patch is notched superiorly and less extended. The stigmata are completely defined in *T. Lorata,* which is found in Washington Territory, and thus replaces *T. Pudens* in the fauna of the West Coast. The types of this species and *Bombycia Semicircularis* are contained in the collection of Mr. Berthold Neumœgen, who has kindly allowed me to figure them.

Thyatira Lorata, Grote, 'Papilio,' i. p. 75.

3. Harrisimemna Trisignata.

This handsome Moth is found in many localities in the Eastern and Middle States. The curious larva has been reared by my kind friend Mrs. Bridgham at her country residence, Seekonk, Rhode Island. The figure is taken from a specimen presented to me by Mrs. Bridgham. It appears that Dr. Harris's name for this Moth was not published until after his death, and in the mean time the species had been described in the British Museum Lists. The genus, which differs strongly from any other, I have named in memory of Dr. Harris; and the common term for the Moth is "Harris's Zigzag."

Grammophora Trisignata, Walk., C. B. M. Noct. p. 29.
Notodonta Sexguttata, Harr., Ent. Cor. p. 174 (figs.).
Harrisimemna Sexguttata, Grote, Tr. Am. Ent. Soc. iv. p. 293.

4. Mastiphanes Edolata.

This species, collected by Mr. Doll in Arizona,

belongs to a section of the genus *Apatela*, characterized by the long and narrow, but hardly pointed primaries, the abdomen well exceeding the hind wings, the thorax elevated. It is allied to the Texan *Extricata*; and probably *Xyliniformis* and *Lithospila* will be found congeneric when the immature stages of all are known. The Californian species described by me as *Spinea* and *Lupini* belong to a different section. The wings are wider, the species more Bombyciform, the thorax globose, the head sunken. I have recently indicated the sections into which the North-American *Apatelæ* seem to fall. I propose to designate the section of which *Spinea* is type by the name *Meroloncbe*.

Apatela Edolata, Grote, ' Papilio,' i. p. 153.

5. Eulonche Lanceolaria.

This interesting form is allied to *E. Oblinita* of Abbot and Smith. It differs in both sexes by the narrower, more lanceolate primaries. It is more bluish than Abbot's species, with less markings; but the transverse posterior line is to be made out, continuous and angulated, not fragmentary and sinuate as in its ally. The hind wings are pure white and have no discal marks. My friend Mr. Roland Thaxter has collected this species in June in Massachusetts. To me it is in many respects the most interesting of all the *Apatelæ*. The section *Eulonche* contains *Oblinita*, *Lanceolaria*, and *Insolita*; it is characterized by the long, pointed, almost Notodontiform primaries and sunken head;

the hind wings have the cell open, vein 5 a mere fold (B. B. S. N. S. i. p. 81).

Eulonche Lanceolaria, Grote, Proc. A. N. S. P. p. 418 (1875).

6. Chytonix Sensilis.

The species of this genus are allied to *Bryophila*, and have the vestiture of the thorax composed of mixed, in part flattened scales. The dorsal line of the body is tufted; the form is more robust than *Bryophila*; in ornamentation there is a white spot attached to the outer median line. M. Guénée placed one species in *Bryophila*, one in *Apamea*. The species are *Jaspis* and *Palliatricula* (perhaps not distinct) and *Sensilis*. This latter has been collected in Massachusetts by Mr. Thaxter. I have not yet seen the male.

Chytonix Sensilis, Grote, 'Papilio,' i. p. 49.

7. Agrotis Perattenta.

This species is found from Canada over the Eastern and Middle States to the southward. It resembles *Agrotis Sigmoides* in ornamentation, but it is a smaller species, and wants the creamy shading over the costa of primaries above from the base outwardly. I have collected it at Buffalo, N.Y., and in the Katskills in June and July.

Agrotis Perattentus, Grote, Can. Ent. vi. p. 131.

8. Agrotis Conchis.

This fine species has been collected by Professor Snow in Colorado. It is allied in ornamentation to

Bicarnea or *C-nigrum*, while quite different from either. The head and collar are bright ochre, the tegulæ brownish ochre. The stigmata contrast with the blackish primaries, which have an overlying lilac-grey cast. All the tibiæ are spinose. With *Hilliana* and *Milleri*, it is one of our handsomest species in the genus.

Agrotis Conchis, Grote, N.-Am. Ent. i. p. 43; id. Tr. Kan. Acad. Sci. vii. p. 66.

9. Agrotis Semiclarata.

This species, of which I figure the male type, has been collected by Mr. H. K. Morrison in Washington Territory. It is allied to *Vancouverensis* and *Gravis* from the West Coast. The species is noteworthy from the sharp division, on the hind wings beneath, of the brownish costal region and the white inferior portion of the under surface, on which latter there is no darker border. The Eastern allies of these three Western species appear to be *Volubilis*, *Stigmosa*, and *Venerabilis*.

Agrotis Semiclarata, Grote, Can. Ent. viii. p. 132.

10. Agrotis Clodiana.

This species has a resemblance to the group of *Gravis*; it is larger, with longer wings, the markings more effaced. The fore wings are of an odd shade of purplish brown with a tinge of yellowish; the fringes have a narrow pale yellowish line at base. Head and collar russet-brown, the latter with a pale line above a darker shade. The female has the fore wings entirely obscure yellowish brown, with the markings lost. Taken by Mr. H. K. Morrison in Washington Territory.

Agrotis Clodiana, Grote, 'Papilio,' i. p. 76.

11. Agrotis Fernaldi.

This fine *Agrotis* has been collected in Maine by Professor C. H. Fernald, after whom the species is named. It is, perhaps, nearer to *Trabalis* than any other Eastern form; the markings are bolder, and the ground-colour more uniform and even. The specimen figured is a typical one, received by me from my good friend Professor Fernald.

Agrotis Fernaldi, Morrison, Proc. A. N. S. Phil. p. 429 (1875).

12. Agrotis Biclavis.

This species has more pointed primaries, and is more distinctly marked than some of the other whitish Western species of *Agrotis*. The veins are inconspicuously marked in white. The distal stigmata are fused, the lines obliterate; there is a fine black basal dash extending into the long claviform spot. The hind wings are pure white in the male, somewhat soiled in the opposite sex. This species, kindly given by Mr. Hy. Edwards, was collected in Arizona. My figure is taken from the female. In the male the antennæ are brush-like, with the joints prominent. All the tibiæ are spinose.

Agrotis Biclavis, Grote, Bull. U. S. Geol. Surv. v. p. 206.

13. Agrotis Parentalis.

This handsome species has been collected in Colorado by Professor F. H. Snow, who has done so much to add to our knowledge of Western Moths. It is allied to another species described by me under the name *Decipiens*; and both have a dorsal elevated ridge of scales on the thorax, which is a character of Lederer's genus *Ammoconia*. Otherwise the structure is like *Agrotis*. The present species differs by the distinct black median lines and reddish shading of the primaries above.

Agrotis Parentalis, Grote, N.-Am. Ent. i. p. 44; id. Tr. Kan. Acad. Sci. vii. p. 66.

14. **Agrotis Specialis.**

This is a richly coloured, red-brown species, which, from the pattern of ornamentation, would be referred by some entomologists to *Ochropleura*, a genus which does not seem to possess any structural characters to separate it from *Agrotis*. The present species is from California, sent me by my kind correspondent Mr. James Behrens, who has, I believe, reared it from the larva. Its nearest ally is *Agrotis Wilsonii,* an olive-coloured species described by me from the same locality.

Agrotis Specialis, Grote, Bull. Buffalo Soc. N. S. ii. p. 62.

15. **Agrotis Vocalis.**

This species has iron-grey fore wings; the median lines are black, thick, single, dentate, the subterminal obsolete. Orbicular spot decumbent, pyriform; reniform incomplete. The hind wings are whitish, with soiled veins and a faint terminal line. The species is found in Colorado.

Agrotis Vocalis, Grote, Can. Ent. xi. p. 56.

16. **Agrotis Pluralis.**

This interesting species is grey along the margins of the fore wings, which are diffusely shaded over the middle with ochreous; the cell is filled in with brown; the lines are double, marked on costa, else merely indicated. The veins are marked with blackish; veins 3 and 4 edged with whitish; a dark shade before subterminal line, resolved into dashes between veins 2 and 5. Hind wings smoky; thorax grey, collar and tegulæ shaded with ochreous. The species has been sent me from Nevada by Dr. James S. Bailey.

Agrotis Pluralis, Grote, Bull. U. S. Geol. Surv. iv. p. 174.

17. **Polia Theodori.**

This fine species, from Colorado, is almost white, shaded with red over the primaries. The tibiæ are unarmed; the eyes naked. In the shape of the thorax it is allied to *Epichysis* and *Aedon*, and together the three seem to form a group of *Polia*. I have described another white Californian species, which, however, has the W-mark more prominent, as *Hadena Olorina*, which resembles the species of this group somewhat, but appears to me to be a true *Hadena*. This species is named after Mr. Theodore S. Bailey.

Apatela Theodori, Grote, Can. Ent. x. p. 237.
Polia Theodori, Grote, Can. Ent. xii. p. 219.

18. **Polia Epichysis.**

This species is purple over grey, with the markings after the pattern of *Theodori*, than which it is slenderer. It was taken by my kind friend Mr. James Behrens, at Soda Springs, California.

Polia Epichysis, Grote, Can. Ent. xii. p. 219.

19. **Heliophila Dia.**

This comparatively short-winged species differs from our Eastern forms by the absence of dark shades accompanying the median vein and other comparative characters. A number of fresh examples have been received from San Francisco by Dr. Bailey.

Heliophila Dia, Grote, Can. Ent. xi. p. 29.

20. **Zosteropoda Hirtipes.**

This singular genus is allied to *Heliophila*. The eyes are naked. The labial palpi rather long; the middle and hind legs tufted. The fore wings are narrow, with parallel margins; the hind wings have a fringing of longer scales on their upper surface along the main nervures. The abdomen is without dorsal tufts, and well exceeds the secondaries. The markings are very simple, and the colour is a brownish orange-yellow. The species has been collected by several entomologists about San Francisco, and was originally sent to me by Mr. Henry Edwards.

Zosteropoda Hirtipes, Grote, Bull. Buff. S. N. S. ii. p. 68, May 1874.

21. **Lithophane Querquera.**

This species has the primaries above of a smooth dusky green. The orbicular is incomplete; the reniform is large, with nebulous reddish centre.

Body flattened; abdomen untufted; head and thorax dusky green, the tegulæ lined outside with black, and a black tuft-point at the middle of the thoracic disk. It was first sent me from Missouri by Professor C. V. Riley; since then it has been taken almost everywhere through the Middle and Eastern States and in Ontario; but nowhere is it common. The Moth hibernates, as in other species in this group.

It is incorrect, in my opinion, to call this genus *Xylina*. This latter name, spelled with an "*e*," is first used by Hübner for *Lithoxylea*. The term *Lithophane* is used for *Petrificata* and four other Noctuæ by Hübner in the 'Verzeichniss.' In 1874 I took *Socia* (*Petrificata*) as the type, and referred *Graptolitha* as synonymous or to be used as a subgenus.

Lithophane Querquera, Grote, Sixth Ann. Rep. Peab. Ac. Sc. p. 34.

22. Lithophane Viridipallens.

This apparently very rare species is of a pale, somewhat bluish, grey-green. It is allied to *Querquera*, but differs by the colour, the narrower reniform, and the absence of the black accentuations to the subterminal line. Hind wing fuscous, with whitish fringes, not ruddy as in its ally; beneath with a faint flush. This is a very delicately coloured and ornamented species; it agrees with *Querquera* in having a neat black dot on the middle of the thorax. Several specimens have been taken by my friend Mr. Roland Thaxter, at Cambridge, Massachusetts.

Lithophane Viridipallens, Grote, Bull. U. S. Geol. Surv. iv. p. 180.

23. **Calocampa Cineritia.**

There are three species of this group in North America, strictly congeneric with the European *Vetusta*. Mr. Morrison regards the Eastern *C. Nupera*, Lintner, as representing the European *C. Vetusta* in America (Bull. Buff. S. N. S. ii. p. 195). The present species seems to have the widest range, extending across the Continent. *C. Cineritia* has the thorax black: there are no brown shades on the internal portion of the wing, nor is the terminal space ochraceous, except opposite the disk and as the continuation of the discal shading; the black-ringed ordinary spots are so close as to nearly meet; the orbicular is not represented by superposed dots as in the European species. This species was sent to me first by Mr. Roland Thaxter from his captures in Massachusetts.

Calocampa Cineritia, Grote, Proc. Ac. N. Sci. Phil. p. 210 (1874).

24. **Gortyna Rigida.**

This species has somewhat the colours of *G. Cataphracta*, but the exterior median line is nearly straight; the terminal fields are purple, the rest of the primary light straw-yellow. The hind wings are pale. It has been taken in Illinois and Pennsylvania, and also by Mr. Thaxter in the Eastern States. There is a tuft behind the collar; the eyes are naked; the front smooth. I refer it to *Gortyna*, Hübner (= *Hydrœcia* of Lederer), although, with *Cerina*, it may find a more natural position when we know its immature form.

Gortyna Rigida, Grote, Can. Ent. ix. p. 87.

25. **Gortyna Cerina.**

This handsome insect has the colours of *Xanthia Togata* (*Silago*), but they are transposed, the terminal field of the fore wings being reddish purple. The ground-colour is yellow, with the markings in broken reddish blotches. There is a long sharp tuft behind the collar; the eyes are naked; front smooth. The head is more sunken and the costa not so straight as in *Xanthia* as defined by Lederer. The female is much larger and heavier than the male *Cerina*. This species is found from Kansas to Massachusetts, where Mr. Thaxter collected the male now figured.

Gortyna Cerina, Grote, Proc. Acad. Nat. Sci. Phil. p. 200 (1874).

26. **Chariclea Triangulifer.**

This golden-yellow species was originally referred by me as the type of the genus *Cirrhophanus*. I believe that there is a slight claw concealed by the vestiture of the fore tibiæ, but have not clearly observed it. The antennæ are simple in the male, scaled above, ciliate beneath, with thickened scape. Eyes naked; thorax tufted; front roughened, tuberculate; palpi slender, short; legs unarmed. The projected ochre-yellow lines form two subtriangular fields on the primaries, which are yellow beneath, with the discal field somewhat fuscous. The species is found from Ohio to Missouri, where I first saw it in Professor C. V. Riley's collections. Afterwards it was sent me by several correspondents for determination. I think it has been redescribed by

Mr. Morrison as *Chariclea Pretiosa*; and this determination has led me to believe a separate genus unnecessary.

Cirrhophanus Triangulifer, Grote, Can. Ent. iv. p. 187; id. Proc. Ac. N. Sci. Phil. p. 421 (1875).

27. Chariclea Pernana.

This species has the colour and markings of the European *Delphinii*; but it is hardly more than a third of the size, and the terminal field, outside of the transverse posterior (t. p.) line on primaries, is olivaceous. A number of specimens are in Mr. Neumœgen's collection, taken by Mr. Doll in Arizona. I have not noted the structure of the fore tibiæ. The eyes are naked, the front roundedly projected; the surface of the clypeus shows a circular rim, enclosing a perpendicular protuberance.

Chariclea Pernana, Grote, 'Papilio,' i. p. 155.

28. Annaphila Superba.

This very pretty little species is the only one of the Californian genus *Annaphila* which has crimson hind wings. The type of the genus, *A. Diva* (Bull. Buff. S. N. S. i. plate iv. fig. 14), has them white. The genus is related to *Eustrotia* (*Erastria* of authors), with some resemblances to *Omia* and the Heliothid genera. The ocelli are unusually large, remote from the constricted, naked compound eyes. Antennæ scaled, ciliate beneath; clypeus full, globose, exceeded by the heavily fringed labial palpi. Body-parts slight in comparison with the wide wings. It differs from *Brephos* by the presence of ocelli. The species look like diminutive *Catocalæ*. The hind wings are brightly coloured, in most of the species yellow. They are all Western, from California to Nevada, and are active in the daytime.

Annaphila Superba, Hy. Edwards, Proc. Cal. Ac. Sci.

29. Grotella Sexseriata.

This genus is related to *Euleucyptera* and *Pippona*; the species look a little like Lithosians, from their long, white, shiny wings. The type of the genus is *G. Septempunctata*, Harvey, figured in the Bulletin of the Buff. Soc. N. Sci. ii. plate iii. fig. 1. *G. Sexseriata* has the primaries crossed by two very distinct black irregular lines, while the subterminal line is represented by a series of black spots. Body-parts yellowish white. Front with a circular projection; eyes naked; fore tibiæ furnished with a claw and a terminal spinule; the tarsi are feebly spinose. Collected by Mr. Doll in Arizona.

Grotella Sexseriata, Grote, 'Papilio,' i. p. 155.

30. Spragueia Funeralis.

The tiny and prettily marked Noctuidæ forming this genus differ structurally from the European *Erotyla Sulphuralis* by the neuration of the secondaries—the European genus being 8-veined, the American 7-veined. On the fore wings also (except in *Onagrus*) veins 8 and 7 are not branched. The clypeus is narrow and smooth in the species of *Spragueia*, flat or but slightly bulging. In *Erotyla* the surface is rough, the infra-clypeal plate is prominent, and the front terminates in a wide-lipped protuberance. I have very fully discussed the genus and allied forms in the 'Canadian Entomologist,' xi. pp. 231–238. The present species is from Arizona. The genus is numerous in kinds; eight are known to me, mostly from the Southern States. The genus is named after my friend, the Entomologist, Henry S. Sprague, of Buffalo, N.Y.

Spragueia Funeralis, Grote, 'Papilio,' i. p. 158.

31. **Adonisea Pulchripennis.**

This brilliantly coloured Californian species differs from the other Heliothid forms by the constricted eyes; the fore tibiæ with a longer inner and two outer claws; the middle and hind tibiæ spinose. The fore wings are stained with brilliant reddish purple with blue-shaded median lines. A variety has been described by Mr. H. Edwards, in which the wings are blackish.

Adonisea Pulchripennis, Grote, Bull. Buff. S. N. Sci. ii. p. 220.

32. **Heliosea Pictipennis.**

This slender Californian species has a single inner claw to the fore tibiæ. The fore wings are pale clay-colour, with two vinous lines, the inner arcuate, the outer subsinuate. Costa and fringes vinous purple.

Heliosea Pictipennis, Grote, Bull. Buff. Soc. N. S. ii. p. 220.

33. **Rhododipsa Volupia.**

In *Rhodophora Florida*, Guén., the fore tibiæ are

provided with an outer claw and two inner spines; the joint is also furnished with spinules. In *Rhododipsa Volupia*, from Texas and Colorado, the joint wants the spinules; there is an outer claw and two spines on the inside, followed by a third more slender. The hind and middle tibiæ are armed; palpi a little longer and more closely scaled. The wings differ in shape and pattern. The head and thorax are dark yellow; the fore and hind wings are bright crimson, the median lines propinquitous, irregular, indistinct, yellowish white. Whether this is the species described by Dr. Fitch under the same specific name is doubtful; but the name need not be changed, unless Fitch's species is different and yet belongs to *Rhododipsa*, which is unlikely.

Rhododipsa Volupia, Grote, B. U. S. G. S. iii. p. 797; id. B. B. E. S. iii. p. 47.

34. Lygranthœcia Acutilinea.

In *Lygranthœcia Marginata*, the type of this genus, the eyes are naked. The fore tibiæ have, on the outside, four stout blunt spines, in a decreasing series towards the base of the joint; all the tibiæ are armed. The group needs re-examination, and some of the numerous species should be separated if possible. *L. Acutilinea* is a very distinct species, characterized by the raggedly toothed silvery median lines on the ochre-fuscous primaries. The hind wings are white, with diffuse blackish discal spot and terminal band. I am indebted to Mr. E. L. Graef for my type, which came from Colorado.

Lygranthœcia Acutilinea, Grote, Can. Ent. x. p. 232.

35. Exyra Rolandiana.

This, the most beautiful species of the genus, has been bred by Mr. Roland Thaxter from the larva feeding on the pitcher-plant, in Massachusetts. The genus is parasitic on the species of *Sarracenia*. It differs structurally from *Xanthoptera*, of which the type is *X. Nigrofimbria*, Guén., by the thicker and longer vestiture. In the type, *E. Semicrocea*, Guén., the 12-veined primaries have the accessory cell greatly elongated, veins 8 and 7 united at base, 9 a very short furcation. Hind wings with vein 5 nearly as strong as the rest. Front not elevated, with a slight inferior tubercle. The primaries have the apices blunted. Four species of *Exyra* are known, viz., *Semicrocea*, *Ridingsii*, *Fax*, and *Rolandiana*.

Exyra Rolandiana, Grote, 'Psyche,' ii. p. 38.

36. Fala Ptycophora.

This singular genus has a number of strong characters. By the conformation of the clypeus it is allied to *Stiria* and *Stibadium*. The front is entirely cup-shaped, excavate, raised around the edges like a rim; from the centre a broad wedge-shaped protuberance arises. Tibiæ unarmed; fore tibiæ with a terminal claw. Eyes naked, unlashed. Thorax without tufts. Male antennæ simple. Fore wings grey, shaded with ochreous, the stigmata distinct on the wide median field. Fringes checkered. Aspect of an *Agrotis*. My type is from California.

Fala Ptycophora, Grote, Proc. Ac. N. Sci. Phil. p. 426 (1875).

37. Aedophron Snowi.

The eyes are naked, the tongue stout, the fore tibiæ short, with terminal and lateral claws; hind and middle tibiæ spinose. The wings are shaped as in the European *A. Rhodites*; the American species is stouter and larger, and while the colours are much the same, the roseate hues are less extended. I received the beautiful species from Professor Snow, after whom it is named. It is the only true *Aedophron* yet described from America. A species is named *Aedophron grandis* in some collections; but this is a synonym of *Copablepharon Absidum*, in which the deep-yellow wings are differently shaped, and shaded variously with greyish fuscous parallel with the margins.

Aedophron Snowi, Grote, Proc. Ac. N. Sci. Phil. p. 422 (1875).

38. Scopelosoma Graefiana.

This bright orange-yellow species is very distinctly marked. Like others of its genus, the Moth hibernates. It has been beaten from oak-leaves in the autumn by Mr. Moffat of Kingston, Ontario, and occurs with *S. Ceromatica*; both of these deeply-coloured species are concealed by the tints of the autumn leaves in which they conceal themselves. It has been collected by Mr. E. L. Graef, after whom the species is named, in New Jersey, and is found also in the Eastern States. The genus is numerously represented in North America. A list of the North-American species belonging to the genera from *Orthosia* to *Lithomia* will be found in the Can. Ent. xii. pp. 155 to 157.

Scopelosoma Graefiana, Grote, Bull. Buff. Soc. N. S. ii. p. 69.

39. **Synedoida Sabulosa.**

This handsome Moth belongs to the *Fasciatæ*. It is found in Southern Colorado; and the typical example I figure has been given to me by Mr. B. Neumœgen.

Synedoida Sabulosa, Hy. Edw., 'Papilio,' i. p. 26.

40. **Catocala Abbreviatella.**

This distinctly marked species is smaller than *C. Nuptialis,* Walk. (of which *Myrrha* is a synonym). It appears to be a more Northern and Western species, occurring in Iowa and Kansas with the more recently described *C. Whitneyi,* which latter is nearer to it in some respects than Mr. Walker's species is. *C. Whitneyi,* however, has hoary fore wings, and the ornamentation constantly differs in detail in series of specimens of both forms collected by Professor Snow. *C. Abbreviatella* has smooth, pale brownish-grey primaries, but little shaded, somewhat darker terminally. The transverse anterior line is black, distinct, and outwardly oblique to below median vein opposite the subreniform, thence lost and obliterate. Reniform annulate, with a black spot inferiorly, not black and pyriform as in *Nuptialis.* The transverse posterior line is faint, plainest opposite the disk, encircling the open subreniform, which is distinctly marked inwardly. Hind wings bright yellow. A rather narrow, slightly constricted median black band, expires much before the margin. Hind border not wide, interrupted. Beneath, the median band is aborted superiorly. The species has been collected in Texas by Mr. Belfrage, in Illinois by Mr. Bean, and in Kansas by Professor Snow.

Catocala Abbreviatella, Grote, Tr. Am. Ent. Soc. iv. p. 14 (Jan. 1872).

41. **Catocala Chelidonia.**

This species has dark fuscous-grey, slightly hoary primaries. Reniform shaded with grey; subreniform independent, stained with brown. The subcostal teeth of the transverse posterior line more distinct than in *C. Similis* (= *C. Formula*), which this species somewhat resembles. Hind wings dark yellow. Median band rather narrow, swollen on the disk, angulated opposite the interruption of marginal band, not reaching internal margin. A number of coinciding examples of this species have been taken by Mr. Doll in Arizona, and are in Mr. B. Neumœgen's collection, to which I am indebted for my types.

Catocala Chelidonia, Grote, 'Papilio,' i. p. 159.

42. **Catocala Beaniana.**

This fine species has been collected in Illinois by Mr. Thos. E. Bean, after whom it is named. It is allied to *C. Meskei* and *C. Briseis*. It is perhaps a little larger than *C. Briseis*; the primaries have a different and paler tint, the transverse posterior line is more dentate, the brown subterminal space paler; the subreniform spot open. Hind wings pinkish red, like *Meskei*; the middle band narrower than in *Briseis*, and more transverse, less rounded.

Catocala Beaniana, Grote, Can. Ent. x. p. 195.
Catocala Beaniana, Hy. Edw., Bull. B. E. S. iii. p. 55.

In 'Papilio,' i. p. 164, I gave a list of the North-American species of *Catocala* described by myself, from which the present form was accidentally omitted. Almost all of these were published by me some time ago, and are now recognized generally in collections. The species of this genus have attracted the attention of collectors from their gay colours and usually large size; and much has been published upon them since the date (1872) of my general paper on our species in the Transactions of the American Entomological Society. Of a few of the following kinds the opinion has been expressed that they are "varieties." It is now nearly twenty years since I described *Catocala Piatrix*; and since that time I have examined a very large material in the genus, so that I am reasonably confident that the following list contains none but valid species, constantly recurring forms which do not intergrade with others. At the outside there are but two or three of which I have not seen many specimens, as, for instance, *Dulciola*, of which, however, Mr. Pilate assures me he has taken a certain number of perfectly constant examples. *C. Alabamæ* is a Southern form, allied to, but perfectly distinct from, *Grynea*, and which I have recently seen again in a collection made in Florida, and thus feel confident that the species is valid. I illustrate here two other species, which have been considered varieties, but never by myself; and it is probable, as proved in the case of *C. Cœlebs*, that the writers who discussed these species had really never seen them, and must have been influenced by other than scientific motives in what they published upon them. The list of species I have described is as follows:—*Retecta, Robinsonii, Flebilis, Levettei, Residua, Angusii, Arizonæ, Meskei, Beaniana, Semirelicta, Coccinata, Verrilliana, Subnata, Piatrix, Adoptiva, Habilis, Cœlebs, Badia, Anna, Clintonii, Abbreviatella, Frederici, Mira, Alabamæ, Præclara, Dulciola, Fratercula, Chelidonia*—in all 28 species.

The species of *Catocala* may be arranged in groups, from the colour of the hind wings. I have arranged them in this way, giving the sections the names used by Hübner in the 'Verzeichniss,' in a paper contributed to the Sixth Annual Report of the Peabody Academy of Sciences, Salem, Mass. The black-winged species (*Mormonia* of Hübner) should inaugurate the genus, and fall into two series, from the colour of the fringes to the secondaries, which are either white or dusky. The orange- and yellow-winged species can hardly be kept separate, as the colour of the hind wings varies to this extent in a single species. In the colour of the collar and fore wings, *C. Concumbens*, a species with lovely pink hind wings, resembles *C. Illecta*, a species with golden-yellow secondaries; the little *C. Tristis*, the smallest species with black hind wings, rather resembles some of the smaller yellow-winged forms, such as *C. Gracilis,* than those of its own group. The type of the genus is the European *C. Fraxini,* which is represented in the North-American fauna by *C. Relicta*; the two agree in the comparative proportion of the body and wings, the body vestiture, cut of wing, colour and ornamentation. That they have a common origin is suggested by the survival of a blue edging to the white band in some specimens of *C. Relicta.*

43. **Catocala Mira.**

This form belongs to the series of *Polygama, Cratægi,* and *Pretiosa.* It is perhaps a little larger, but in colour of primaries nearest to *Polygama,* from which it differs by the absence of any determinate greenish or brown and pale shades on the more uniform fore wings. The lines are more denticulate than in the other species. The hind wings are of a deeper, almost orange-yellow, and at base and along internal margin noticeably free from dusky hair and scales. Professor Snow has collected this species commonly at Lawrence, Kansas.

Catocala Mira, Grote, Can. Ent. viii. p. 230.

44. **Catocala Frederici.**

This species has clouded pale greenish-grey fore wings, quite unlike any other described species. The hind wings are largely clear bright yellow. A narrow median black band, tapering inferiorly, and discontinued shortly beyond its inward projection, expiring much before internal margin; terminal band narrow, abruptly discontinued at vein 2; a small spot at anal angle. Beneath, both wings clear yellowish, with narrow bands. This very distinct species was first collected in Southern Texas by Friederich, after whom the species is named. My types are in the Imperial Museum at Berlin. The species was afterwards re-discovered by Belfrage.

Catocala Frederici, Grote, Trans. Am. Ent. Soc. p. 14 (1872).

45. Strenoloma Lunilinea.

This fine Moth has been found in Virginia, and also in Ohio, where Mr. Dury collected it about Cincinnati in some number. Professor Snow has also taken it in Kansas; and my figure is drawn after a male specimen furnished by him. The Moth has delicate dove-coloured grey wings; the costal edge is marked by four distinct brown liturate spots, marking the inception of the subobsolete transverse lines. The collar is more brownish and darker than the body and wings. On the primaries the exterior margin is a little darker shaded, often in the males with a shade-spot about the middle. My original specimen was collected by Mr. Theo. L. Mead. The middle tibiæ are sparsely spinose.

Spiloloma Lunilinea, Grote, Bull. B. S. N. S. i. p. 127.
Strenoloma Lunilinea, Grote, N. Am. Ent. i. p. 99.

A.H. Searle, del et lith. Mintern Bro's imp

1. Semicircularis. 2. Lorata. 3. Trisignata. 4. Edolata.
5. Lanceolaria. 6. Sensilis. 7. Perattenta. 8. Conchis.
9. Semiclarata. 10. Clodiana. 11. Fernaldi. 12. Biclavis. 13. Parentalis.

PLATE I.

Fig. 1. Bombycia Semicircularis.
2. Thyatira Lorata.
3. Harrisimemna Trisignata.
4. Mastiphanes Edolata.
5. Eulonche Lanceolaria.
6. Chytonix Sensilis.
7. Agrotis Perattenta.
8. —— Conchis.
9. —— Semiclarata.
10. —— Clodiana.
11. —— Fernaldi.
12. —— Biclavis.
13. —— Parentalis.

A.H. Searle, del et lith. Mintnern Bro's imp.

14. Specialis. 15. Vocalis. 16. Pluralis. 17. Theodori.
18. Epichysis. 19. Dia. 20. Hirtipes. 21. Querquera.
22. Viridipallens. 23. Cineritia. 24. Rigida. 25. Cerina.

PLATE II.

Fig. 14. Agrotis Specialis.
15. —— Vocalis.
16. —— Pluralis.
17. Polia Theodori.
18. —— Epichysis.
19. Heliophila Dia.
20. Zosteropoda Hirtipes.
21. Lithophane Querquera.
22. —— Viridipallens.
23. Calocampa Cineritia.
24. Gortyna Rigida.
25. —— Cerina.

A.H. Searle, del. et lith. Mintern Bro's imp.

26. Triangulifer. 27. Pernana. 28. Superba. 29. Sexseriata. 30. Funeralis. 31. Pulchripennis. 32. Pictipennis. 33. Volupia. 34. Acutilinea 35. Rolandiana. 36. Ptycophora. 37. Snowi. 38. Graefiana.

PLATE III.

Fig. 26. Chariclea Triangulifer.
27. —— Pernana.
28. Annaphila Superba.
29. Grotella Sexseriata.
30. Spragueia Funeralis.
31. Adonisea Pulchripennis.
32. Heliosea Pictipennis.
33. Rhododipsa Volupia.
34. Lygranthœcia Acutilinea.
35. Exyra Rolandiana.
36. Fala Ptycophora.
37. Aedophron Snowi.
38. Scopelosoma Graefiana.

39.

40.

41.

42.

43.

44.

45.

A.H. Searle, del. et lith. Mintern Bro's imp.

39. Sabulosa. 40. Abbreviatella. 41. Chelidonia. 42. Beaniana.
43. Mira. 44. Frederici. 45. Lunilinea.

PLATE IV.

Fig. 39. Synedoida Sabulosa.
40. Catocala Abbreviatella.
41. —— Chelidonia.
42. —— Beaniana.
43. —— Mira.
44. —— Frederici.
45. Strenoloma Lunilinea.

A COLONY OF BUTTERFLIES.

BY

AUGUSTUS R. GROTE.

TO THE

Memory

OF

THOMAS SAY.

A COLONY

OF

BUTTERFLIES.

About one hundred thousand years ago, during the decline of the Ice period, a colony of Butterflies settled in New England. They chose for their territory Mount Washington, in New Hampshire, and their descendants occupy the rocky summit of that mountain to this day.

Mount Washington is 6293 feet high, and the White-Mountain butterflies are not found below an elevation of about 5600 feet. Between this height and the often cloud-capped summit, the butterflies disport during the month of July of every year. The bare and inhospitable mountain-top affords little vegetation, yet the White-Mountain butterflies find there food upon which they thrive. Both Mr. Sanborn and Mr. Scudder have found the caterpillar feeding upon the sedges, which grow, as best they may, in hollows and between the rocks. The dusky brown butterfly, which succeeds the caterpillar, measures about one and eight tenths inches from tip to tip of the extended front wings. Above, the four wings are feebly marked; beneath, the hind pair are crossed by a dark median band with its outer edges deeper brown and irregular, while beyond the band the wings are marbled, brown and white.

Naturalists know the White-Mountain butterfly by the name of *Oeneis Semidea,* and its first biographer was Thomas Say, who described it in the year 1828. Previously, Mr. Thomas Nuttall, the botanist, had collected specimens of the butterfly, while Say's original figure of the species was drawn from an individual presented to him by Mr. Charles Pickering, of Salem.

It is 1800 miles west from Mount Washington to Long's Peak, Colorado. In this direction, over all the level stretch of country, no butterflies like our White-Mountain butterfly are to be met with. But, in Colorado, species similar to the White-Mountain butterfly, and probably one exactly like it, are found again occupying elevated lands. To the northward it is 1000 miles to Hopedale, Labrador, and here again very similar butterflies are found living in that barren region.

This is a strange distribution for a butterfly, and so the question comes up as to the manner in which it was brought about. By comparing what has been found out, with regard to past conditions of the earth and the present state of things, a solution of the question has been offered. This solution gives us the Ice period in North America as the agent, which induced the present distribution of the genus to which the White-Mountain butterfly belongs. And the colonization of the butterfly, on our New-England mountains, would have been effected in this wise.

Before the Ice reign commenced in New England, it had extended itself over the north of the continent. The ice gradually and very slowly advanced, year by year, to the southward. Always more snow fell than was melted, and this snow stayed, summer and winter, and accumulated more and more. It consolidated into *névé* and glacial ice. Forming on the highest lands, the ice-rivers filled the ravines and joined, upon the plains, the main body of ice which was pressing southward from the pole. Summer and winter still alternated, but, as is the case now in the extreme north, the summers were short and the winters long. The advancing ice destroyed, or drove before it, the insects and animals of the warmer climates, which it chilled by its approach. But it was kind to its own children. It brought down with it the *Oeneis* butterflies and the reindeer. Before its feet it spread food for both of these, year by year, always pushing food and animals to the south. At the probable rate of less than a mile in a hundred years, it brought them at last from the farthest north into Virginia; not the Virginia of to-day, but Virginia changed into an Arctic scene*.

At length the climate began again to change. The point of farthest advance reached, the ice commenced to retrace its steps. And it called its own back with it, alluring them by their food, scattered ever farther and farther to the north. At some time, the lengthening summers and shortening winters brought the main Ice-sheet back into New England. From Southern New York to Connecticut, to Massachusetts, to Vermont, to New Hamp-

* Consult A. R. Grote, 'Proceedings of the American Association for the Advancement of Science,' p. 222 (1875); also 'Silliman's Journal' for the same year.

shire, it retreated all the way. It was as the falling back of an army, with all its baggage and equipments, and in perfect order. Year by year it called upon its plants, its butterflies, its animals, and they followed in its royal train. It had overridden all obstacles, all lives and constitutions, and in its retreat it shed, over the lands which again saw the sun, floods of water, the source of fresh life and civilizations.

But it was careful of its own plants and animals; they were to go back with the ice, nor be seduced by the lakes and streams its retreat unveiled, and so become companions to the mammoth. And it succeeded, for the most part, until it reached the White Mountains. Though, year by year, the individual butterflies perished, they planted their successors; the longer-lived reindeers laid their bones by the way, and in the Connecticut Valley itself, but fresh herds still were ready to follow the northward march of the great glacier.

Out of the valley of the White Mountains, the main ice-mass gradually retreated; and here it lost some of its followers. At that time the White Mountains must have presented an appearance not unlike the Alps of to-day—an aspect which, owing to their inferior elevation, they have since lost under a climate growing in warmth. The local glaciers, which then filled the mountain-gullies, attracted some of the wayward, flitting *Oeneis* butterflies by a display of the food-plants which they had harboured and detained from the main glacier.

Year after year the great glacier retreated farther

and farther north, followed by the main body of its train—plants, butterflies, and animals,—the while some of these foolish butterflies were beguiled by the shallow ice-torrents which then filled the ravines of Mount Washington. Return became at length impossible. They advanced behind the deceiving local glaciers, step by step up the mountain-side, pushed from below by the warm climate, which to them was uncongenial, until they reached the mountain-peak, to-day bare of snow in the short summer. Here, blown sideways by the wind, they patiently cling to the rocks. Or, in clear weather, on weak and careful wing, they fly from flower of stemless mountain-pink to blue-berry, swaying from their narrow tenure of the land. Drawn into the currents of air that sweep the mountain's side, they are, at times, forced downwards, to be parched in the hot valleys below. Yet they maintain themselves. They are fighting it out on that line. They are entrapped, and must die out by natural causes, unless certain entomologists sooner extirpate them by pinning them up in collections of insects.

What time, on "Bigelow's Lawn," I see the ill-advised collector, net in hand, swooping down on this devoted colony, of ancient lineage and more than Puritan affiliation, I wonder if, before it is too late, there will not be a law passed to protect the butterflies from the cupidity of their pursuers.

This is the story of a New-England colony of butterflies. I commend this colony to the protection of all good citizens of the State of New Hampshire.